OXFORD BOOKWORMS LIBRARY

Human Interest

The Garden Party
and Other Stories

Stage 5 (1800 headwords)

Series Editor: Jennifer Bassett
Founder Editor: Tricia Hedge
Activities Editors: Jennifer Bassett and Alison Baxter

KATHERINE MANSFIELD

The Garden Party

and Other Stories

Retold by
Rosalie Kerr

OXFORD UNIVERSITY PRESS

Oxford University Press
Great Clarendon Street, Oxford OX2 6DP

Oxford New York
Athens Auckland Bangkok Bogotá Buenos Aires Cape Town
Chennai Dar es Salaam Delhi Florence Hong Kong Istanbul Karachi Kolkata
Kuala Lumpur Madrid Melbourne Mexico City Mumbai Nairobi
Paris São Paulo Shanghai Singapore Taipei Tokyo Toronto Warsaw
with associated companies in
Berlin Ibadan

OXFORD and OXFORD ENGLISH
are trade marks of Oxford University Press

ISBN 0 19 423065 1

This simplified edition © Oxford University Press 2000

Third impression 2001

First published in Oxford Bookworms 1998
This second edition published in the Oxford Bookworms Library 2000

Illustrated by Susan Scott

Printed in Spain by Unigraf s.l.

CONTENTS

Feuille d'Album

He really was an impossible person. Too shy, and he had nothing at all to say. When he came to your studio, he just sat there, silent. When he finally went, blushing red all over his face, you wanted to scream and throw something at him.

The strange thing was that at first sight he looked most interesting. Everybody agreed about that. You saw him in a café one evening, sitting in a corner with a glass of coffee in front of him. He was a thin, dark boy, who always wore a blue shirt and a grey jacket that was a little too small for him. He looked just like a boy who has decided to run away to sea. You expected him to get up at any moment, and walk out into the night and be drowned.

He had short black hair, grey eyes, white skin and a mouth that always looked ready for tears. Oh, just to see him did something to your heart! And he had this habit of blushing. If a waiter spoke to him, he turned red!

'Who is he, my dear? Do you know?'

'Yes. His name is Ian French. He paints. They say he's very clever. Someone I know tried to mother him. She asked him how often he had a letter from home, if he had enough blankets on his bed, how much milk he drank. Then she went to his studio to make sure he had enough clean shirts. She rang and rang the bell, but nobody came to the door, although she was sure he was there … Hopeless!'

1

Someone else decided he ought to fall in love. She called him to her, took his hand, and told him how wonderful life can be for those who are brave. But when she went to his studio one evening, she rang and rang ... Hopeless.

'What the poor boy really needs is excitement,' a third woman said. She took him to cafés and night-clubs, dark places where the drinks cost too much and there were always stories of a shooting the night before. Once he got very drunk, but still he said nothing, and when she took him home to his studio, he just said 'goodnight' and left her outside in the street ... Hopeless.

Other women tried to help him – women can be *very* kind – but finally they, too, were defeated. We are all busy people, and why should we spend our valuable time on someone who refuses to be helped?

'And anyway, I think there is something rather odd about him, don't you agree? He can't be as innocent as he looks. Why come to Paris if you don't intend to have any fun?'

He lived at the top of a tall, ugly building, near the river. As it was so high, the studio had a wonderful view. From the two big windows he could see boats on the river and an island covered with trees. From the side window he looked across to a smaller and uglier house, and down below there was a flower market. You could see the tops of huge umbrellas with bright flowers around them, and plants in boxes. Old women moved backwards and forwards among the flowers. Really, he didn't need to go out. There was always something to draw.

If any kind woman had been able to get into his studio, she would have had a surprise. He kept it as neat as a pin. Everything was arranged in its place, exactly like a painting – the bowl of eggs, the cups and the teapot on the shelf, the books and the lamp on the table. There was a red Indian cover on his bed, and on the wall by the bed there was a small, neatly written notice: GET UP AT ONCE.

Every day was the same. When the light was good he painted, then cooked a meal and tidied the studio. In the evenings he went to the café or sat at home reading or writing a list which began: 'What I can afford to spend'. The list ended 'I promise not to spend more this month. Signed, Ian French.'

Nothing odd about that; but the women were right. There was something else.

One evening he was sitting at the side window eating an apple and looking down on to the tops of the huge umbrellas in the empty flower market. It had been raining, the first spring rain of the year, and the air smelled of plants and wet earth. Down below in the market, the trees were covered in new green. 'What kind of trees are they?' he wondered. He stared down at the small ugly house, and suddenly two windows opened like wings and a girl came out on to the balcony, carrying a pot of daffodils. She was a strangely thin girl in a dark dress, with a pink handkerchief tied over her hair.

'Yes, it is warm enough. It will do them good,' she said, putting down the pot, and turning to someone in the room inside. As she turned, she put her hands up to her hair to tidy it, and looked down at the market and up at the sky. She did not look at the house opposite. Then she disappeared.

His heart fell out of the window and down to the balcony, where it buried itself among the green leaves of the daffodils.

The room with the balcony was the sitting-room, and next to it was the kitchen. He heard her washing the dishes after supper, saw her come to the window to shake out the tablecloth. She

3

never sang or combed her hair or stared at the moon as young girls are said to do. She always wore the same dark dress and pink handkerchief.

Who did she live with? Nobody else came to the window, but she was always talking to someone. Her mother, he decided, was always ill. They took in sewing work. The father was dead ... He had been a journalist. By working all day she and her mother just made enough money to live on, but they never went out and they had no friends.

He had to make some new notices ... 'Not to go to the window before six o'clock: signed, Ian French. Not to think about her until he had finished his painting for the day: signed, Ian French.'

It was quite simple. She was the only person he wanted to know because she was, he decided, the only person alive who was exactly his age. He didn't want silly girls, and he had no use for older women. She was his age. She was – well, just like him.

He sat in his studio, staring at her windows, seeing himself in those rooms with her. She was often angry. They had terrible fights, he and she. And she rarely laughed. Only sometimes, when she told him about a funny little cat she once had, who used to scratch and pretend to be fierce when she gave it meat to eat ... Things like that made her laugh. Usually, they sat together very quietly, talking in low voices, or silent and tired after the day's work. Of course, she never asked him about his pictures, and of course he painted the most wonderful pictures of her, which she hated because he made her so thin and so dark ...

But how could he meet her?

Then he discovered that once a week, in the evening, she went shopping. On two Thursdays he saw her at the window in a coat, carrying a basket. The next Thursday, at the same time, he ran down the stairs. There was a lovely pink light over everything. He saw it reflected in the river, and the people walking towards him in the street had pink faces and pink hands.

Outside the house he waited for her. He had no idea what he was going to do or say. 'Here she comes,' said a voice in his head.

Outside the house he waited for her.

She walked very quickly, with small, light steps … What could he do? He could only follow …

First she went to buy some bread. Then she went to a fish shop. She had to wait a long time in there. Then she went to the fruit shop and bought an orange. As he watched her, he knew more surely than ever that he must talk to her, now. Her seriousness and her loneliness, even the way she walked – separate, somehow, distant from the other people in the street – all this was so natural, so right to him.

'Yes, she is always like that,' he thought proudly. 'She and I are different from these people.'

But now she was going home, and he had not spoken to her. Then she went into another shop. Through the window, he saw her buying an egg. She took it carefully out of the basket – a brown egg, a beautiful one, the one he himself would have chosen. She came out of the shop, and he went in. A moment later he was out again, following her through the flower market, past the huge umbrellas, walking on fallen flowers.

He followed her into the house and up the stairs. She stopped at a door and took a key out of her purse. As she put the key in the lock, he ran up to her.

Blushing redder than ever, but looking straight at her, he said, almost angrily: 'Excuse me, Mademoiselle, you dropped this.'

And he gave her an egg.

The Doll's House

When dear old Mrs Hay went home after staying with the Burnells, she sent the children a doll's house. It was so big that Pat, the hired man, could only just lift it, and they had to leave it outside in the garden. It was all right there; it was summer. And perhaps the smell of paint would go before they had to take it indoors. Really, the smell of paint (so sweet of dear, generous Mrs Hay!) – but the smell of paint was so strong that it was enough to make anyone seriously ill, or so Aunt Beryl thought. You could smell the paint even when it was wrapped up. And when they unwrapped it …

There it stood, a doll's house, painted a thick, dark, oily green. Its two solid little chimneys were painted red and white, and it had a bright yellow door and real glass windows.

It was perfect! Who cared about the smell? It was part of the wonder of the doll's house, part of the newness.

'Open it quickly, someone!'

The fastening at the side was stuck fast, and Pat had to use his knife to get it open. But then … the whole front of the house swung back and – you could see everything! The sitting-room, the kitchen, the two bedrooms. That is the way for a house to open! Why don't all houses do that? How exciting, to be able to see everything you want to see, all at once!

'Oh – oh!' The Burnell children were quite unable to speak. It was too wonderful. They had never seen anything like it in their lives. There was paper on the walls, and pictures, just like in real houses. There was red carpet on the floors, except in the kitchen; red and green chairs, beds with real covers on them, tiny plates and cups.

But what Kezia liked more than anything, what she liked most awfully, was the lamp. It stood in the middle of the table, a beautiful little gold and white lamp, all ready to be lit. Of course, you couldn't really light it, but there was something inside it that looked like oil and moved when you shook it.

The mother and father dolls sitting stiffly in their chairs and their two little children in bed upstairs were really too big for the doll's house. They didn't look quite right. But the lamp was perfect. It seemed to smile at Kezia, to say, 'I live here.' The lamp was real.

❧

The Burnell children could not walk to school fast enough the next morning. They burned to tell everybody about the doll's house.

'I shall describe it,' said Isabel, 'because I'm the eldest. You two can join in, but I must speak first.'

Lottie and Kezia said nothing. Isabel was bossy, but she was always right.

'And I shall choose who's going to come and see it first,' Isabel said. 'Mother said I could.'

Their mother had told them that they could ask the girls at school, two at a time, to come and see the doll's house. Of course, they were not invited to tea, or to come into the house, but they could come into the garden and stand there quietly, while Isabel showed them all the lovely things in the doll's house.

It was too bad that they arrived at school just as the bell was ringing, and they had no time to talk to anyone. Never mind! Isabel looked very important and mysterious, and whispered to some of her friends, 'I've got something to tell you at play-time!'

When play-time came, everyone wanted to be near Isabel. The little girls almost fought to put their arms around her, to walk beside her and be her special friend. Laughing and pushing one another, they gathered closely around her. The only two who stayed outside the circle were the two who were always outside – the Kelveys. They knew they were not wanted.

To be perfectly honest, the school the Burnell children went to was not the kind of school their parents really wanted for them. But they had no choice. It was the only school for miles. And because of this, all the children in the area, the Judge's little girls, the doctor's daughters and all the children of milkmen and farmers, were forced to mix together. And there were plenty of rude, rough little boys, too. But worst of all, there were the Kelveys. The Burnell children were not allowed to speak to them. They walked past the Kelveys with their heads in the air. And because others followed where the Burnells led, nobody spoke to the Kelveys. Even the teacher had a special voice for them, and a special smile for the other children when Lil Kelvey came up to her desk to give her some very tired-looking flowers she had picked by the side of the road.

They were the daughters of a neat, hard-working little woman, who went from house to house, doing people's washing for them. This was awful enough. But where was Mr Kelvey? Nobody knew. But everybody said he was in prison. So they were the daughters of a woman who washed people's clothes and a man who was in prison. Very nice companions for other people's children!

Then there was the way they looked. It was hard to understand why Mrs Kelvey dressed them in such an odd way. The truth was, she made their clothes from old bits and pieces which were given to her by the people she worked for. Lil, for example, who was a solid, plain child, came to school in a dress made out of an old green tablecloth of the Burnells, and a red curtain that had belonged to the Logans. Her hat came from Miss Lecky at the post office, and had a long red feather stuck in it. What a sight she looked! It was impossible not to laugh.

And her little sister, 'our Else', as Lil always called her, wore a long white dress that looked like a night-dress, and a pair of boy's boots. But our Else would have looked strange in any clothes. She was a tiny white creature with huge eyes – just like a little bird. Nobody had ever seen her smile; she hardly ever spoke. Everywhere Lil went, our Else followed, holding a piece of Lil's skirt in her hand. In the playground or on the road to or from school, you could always see Lil, with our Else close behind her. When she wanted something, our Else pulled on Lil's skirt, and Lil stopped and turned around. The Kelveys always understood one another.

Now they stood at the edge of the circle, outside the group of friends. You couldn't stop them listening. When the little girls turned round and gave them angry looks, Lil, as usual, smiled all over her silly red face, but our Else just stared and said nothing.

Isabel talked on, in a proud voice. She described the carpet, the beds with real covers, the kitchen with its tiny cups and plates.

When she finished, Kezia said, 'You've forgotten the lamp, Isabel.'

'Oh yes,' said Isabel. 'There's a lovely little lamp on the table. It's just like a real one.'

'The lamp's best of all,' cried Kezia. She wanted Isabel to talk for longer about the lamp, to let everyone know how special it was. But nobody was listening to Kezia. Isabel was choosing the first two who were going to come and see the doll's house. She chose Emmie Cole and Lena Logan. But all the others knew that they would have a chance to see it another day. They all wanted to be nice to Isabel. They all had a secret to whisper to her. 'Isabel's *my* friend.'

Only the little Kelveys were forgotten. There was nothing more for them to hear.

Days passed, and more and more children were taken to see the doll's house. It was the one thing they talked about. 'Have you seen the Burnells' doll's house? Oh, isn't it lovely? Haven't you seen it yet? Oh, dear!'

The little girls talked about the doll's house at dinner-time, as they sat under the trees in the school playground, eating their thick meat sandwiches and buttered cake. The little Kelveys listened, while they ate their bread and butter out of a piece of newspaper.

'Mother,' said Kezia, 'please can I ask the Kelveys, just once?'

'Of course not, Kezia.'

'But why not?'

'Run away, Kezia. You know why not.'

At last the day came when everyone except the Kelveys had seen the doll's house. That day, there was less to talk about. It was dinner-time. The little girls sat together under the trees, and suddenly, as they looked at the Kelveys eating out of their piece of newspaper, they wanted to be unkind to them.

Emmie Cole started it. 'Lil Kelvey's going to be a servant when she grows up,' she whispered.

'Oh, how awful!' said Isabel Burnell.

Emmie looked at Isabel in a way she had seen her mother look, when she was talking about things like this.

'It's true,' she said.

Then Lena Logan joined in. 'Shall I ask her?' she said.

'You don't dare,' said Jessie May.

'Oh, I'm not frightened,' said Lena. She laughed and did a little dance in front of the other girls. 'Watch! Watch me now!' she said, and she danced right over to the Kelveys.

Lil looked up from her bread and butter. Our Else stopped eating. What was coming now?

'Is it true that you're going to be a servant when you grow up, Lil Kelvey?' Lena screamed at her.

Silence. Lil gave no answer, but she smiled her silly, red-faced smile. She didn't seem to mind the question at all. Poor Lena! The other girls began to laugh at her.

Lena didn't like that. She stepped right up to Lil. 'Yah, your father's in prison!' she shouted in her face.

This was so wonderful to hear that all the little girls rushed away together, deeply excited by what Lena had done. How fast they ran, how high they jumped, how wild and free they felt that morning!

In the afternoon, Pat came to take the Burnell children home. There were visitors. Isabel and Lottie, who liked visitors, went upstairs to change their dresses, but Kezia slipped quietly out into the garden. There was nobody there. She began to swing on the big white garden gate. Then, looking down the road, she saw two little figures coming towards her, one in front, the other close behind. It was the Kelveys. She got down from the gate. For a moment she thought about running away. The Kelveys came nearer. Then Kezia climbed back up on the gate. She had decided what she must do. She started swinging on the gate again.

'Hello,' she said to the Kelveys.

They were so surprised that they stopped. Lil gave her silly smile. Our Else stared.

'You can come and see our doll's house if you want to,' Kezia said.

Lil turned red. She shook her head.

'Why not?' asked Kezia.

'Your ma told our ma you mustn't speak to us.'

'Oh, well,' said Kezia. She didn't know what to say. 'It doesn't matter. But you can still come and see our doll's house. Come on. Nobody's looking.'

But Lil shook her head again.

'Don't you want to?' asked Kezia.

Suddenly, there was a pull on Lil's skirt. She turned round. Our Else was looking at her with big, desperate eyes. She wanted to see the doll's house. Lil looked at her very doubtfully. But then our Else pulled her skirt again. Lil stepped forwards. Like two little lost cats, they followed Kezia across the garden to where the doll's house stood.

'There it is,' said Kezia.

They said nothing. Lil breathed loudly. Our Else was as still as stone.

'I'll open it for you,' said Kezia kindly. 'Look, here's the sitting-room and the kitchen, and that's the—'

'Run away, children, and don't come back!' Aunt Beryl said to the Kelveys.

'Kezia!'

Oh, how they jumped!

'Kezia!'

It was Aunt Beryl's voice. They turned round. She was standing at the back door, staring at them. Aunt Beryl just couldn't believe her eyes.

'How dare you bring the little Kelveys into our garden!' she said to Kezia, in a cold, angry voice. 'You know as well as I do that you aren't allowed to talk to them.'

'Run away, children, run away and don't come back!' she said to the Kelveys. 'Off you go immediately!'

She did not have to tell them twice. They were out of the garden in a moment, Lil red-faced and ashamed, with our Else hanging onto her skirt.

'Bad, disobedient little girl!' Aunt Beryl said bitterly to Kezia, and she closed the doll's house with a bang.

Aunt Beryl had been having a terrible day, but now that she had got rid of those little animals the Kelveys and shouted at Kezia, she felt a lot better. She went back into the house singing.

When the Kelveys were far away from the Burnells' house, they stopped and sat down by the side of the road. Lil's face was still burning, and she took off her hat. They stared across the fields, where the Logans' cows were eating grass. What were the little Kelveys thinking?

Our Else moved closer to her sister. She had already forgotten the angry lady. She put out a finger and touched the feather on Lil's hat. She smiled her rare smile.

'I seen the little lamp,' she said softly.

Then both were silent once more.

15

The Garden Party

They need not have worried. The weather was perfect – warm, and without a cloud in the sky. The gardener had been at work since dawn, cutting and brushing the lawns, until the green grass shone. And the roses – the roses were wonderful. Hundreds of flowers had opened during the night. You could almost believe that the roses knew about the garden party!

While the Sheridan girls were having breakfast, the men came with the marquee.

'Where shall we put the marquee, mother?' asked Meg.

'My dear child, please don't ask me. I'm determined to make you children organize everything this year. Forget that I am your mother. Pretend I'm one of your guests.'

But Meg could not possibly talk to the men. She had just washed her hair. Jose, as usual, wasn't even dressed yet.

'You'll have to go, Laura. You're the artistic one in this family.'

Laura flew out of the house, still holding a piece of bread and butter in her hand. Food always tasted delicious out of doors, and Laura loved arranging things. She always felt that she could do it better than anyone else.

Four men were waiting on the garden path. They were carrying big bags of tools, and looked very serious. Laura wished she had left her bread and butter in the house. She blushed, and tried to look business-like.

16

'Good morning,' she said, copying her mother's voice. But it sounded so silly that she was ashamed, and asked, just like a little girl, 'Oh, have you come – is it about the marquee?'

'That's right, miss,' said the tallest of the men. He pushed back his hat and smiled down at her.

His smile was so friendly that Laura felt better immediately. What nice eyes he had – small, but a lovely dark blue! All the men were smiling now. 'Cheer up! We won't bite!' they seemed to be saying. How very nice workmen were! And what a beautiful morning! She mustn't mention the morning; she must be business-like. The marquee.

'Well, shall we put it on the lawn over there?'

She pointed with the hand that was not holding the bread and butter. They all turned and stared. The tall man frowned.

'I don't like it,' he said. 'You wouldn't notice it there. You see, with a thing like a marquee, you want it where it hits you – bang in the eye, as you might say.'

Laura had been brought up in a way which made her wonder for a moment whether a workman should use an expression like 'bang in the eye' to her. But she understood what he meant.

'A corner of the tennis court,' she suggested. 'But the band's going to be in one corner.'

'Having a band, are you?' said another workman. He was pale, with a tired look in his dark eyes. What was he thinking?

'Only a very small band,' Laura said gently. Perhaps he wouldn't mind a very small band. But the tall man said, 'Look here, miss, that's the place. By those trees. Over there.'

By the karaka trees. The marquee would hide them. And the karaka trees were so lovely, with their big, shiny leaves and orange fruit. Must they be hidden by a marquee?

17

They must. The men were already carrying their bags of tools across the lawn. Only the tall man was left. Suddenly, he bent down, touched a rose, and pulled it gently towards him to smell it.

When Laura saw him do that, she forgot about the karakas. He was a workman who loved the perfume of roses. How many of the men that she knew cared about things like that? Oh, how nice workmen are, she thought. Why couldn't she have them for her friends, instead of the silly boys she danced with and who came to Sunday night supper? She liked these men much better.

It's all the fault, she decided, of these stupid differences in social class. Well, for her there *were* no differences. Absolutely none at all, not a single one … And now there came the sound of hammers. Someone whistled, someone called out, 'Are you all right, mate?' 'Mate!' How friendly they were! Just to show how happy she was, how she liked being among these friendly men, Laura took a big bite out of her bread and butter. She felt just like a work-girl.

'Laura, Laura, where are you? Telephone, Laura!' a voice cried from the house.

'Coming!' She ran across the lawn, up the path and into the house. In the hall, her father and Laurie were brushing their hats, getting ready to go to the office.

'I say, Laura,' said Laurie, 'take a look at my coat, can you, before this afternoon? I think it needs ironing.'

'All right,' she said. Suddenly, she couldn't stop herself. She ran up to Laurie and threw her arms around him. 'Oh, I do love parties, don't you!' she cried.

'I'll say I do!' said Laurie's warm, boyish voice. He gave his sister a gentle push. 'Run off to the phone, old girl.'

The telephone. 'Yes, yes; oh yes. Kitty? Good morning, dear.

Come to lunch, my dear! It will be nothing special – just what's left over. Yes, isn't it a perfect morning? Yes, wear your white dress. One moment – mother is saying something.'

Mrs Sheridan's voice floated down the stairs. 'Tell her to wear that sweet hat she wore last Sunday.'

'Mother says you must wear that *sweet* hat you wore last Sunday. Good. One o'clock. Bye-bye!'

Laura put down the phone, took a deep breath, and stretched out her arms. Then she stood still, listening. The house was alive with sounds of running feet and distant voices. Somewhere down in the kitchen, a door opened and closed. Sunlight, and little warm winds, played in and out of the windows. Darling little winds.

The door-bell rang, and she heard a man's voice and then Sadie saying, 'I'm sure I don't know. Wait. I'll ask Mrs Sheridan.'

'What is it, Sadie?' Laura came into the hall.

'The flowers have come from the shop, Miss Laura.'

And there they were, by the door. Box after box, full of pots of pink lilies. No other kind. Nothing but lilies, big pink flowers, wide open and almost frighteningly alive.

'O–oh, Sadie!' said Laura. She bent down to touch them, half expecting the flowers to burn her fingers.

'It must be a mistake,' she said softly. 'Nobody ever ordered so many. Sadie, go and find mother.'

But at that moment Mrs Sheridan appeared.

'It's quite right,' she said calmly. 'I ordered them. Aren't they lovely?' She touched Laura on the arm. 'I was passing the shop yesterday, and I saw them in the window. I thought – for once in my life I shall have enough lilies! The garden party will be a good excuse.'

'But I thought that we children had to organize everything this year,' said Laura. Sadie had gone, and the man from the flower-shop was outside. She put her arm around her mother's neck, and gently, very gently, she bit her mother's ear.

'My darling child, you wouldn't like me to be a sensible mother, would you? Don't do that. Here's the man.'

He was carrying in another box of lilies.

'Put them here, please, on either side of the door,' said Mrs Sheridan. 'Don't you agree that they'll look best there, Laura?'

'Oh, *yes*, mother.'

In the sitting-room, Meg, Jose and little Hans were arranging the furniture.

'Now we should put the piano here, and move everything else except the chairs out of the room, don't you think?'

'Exactly.'

'Hans, move these tables into the smoking-room, and then brush the carpet, and – one moment, Hans.'

Jose loved giving orders to the servants, and they loved obeying her. She made them feel that they were all acting together in some exciting play.

'Tell mother and Miss Laura to come here at once.'

'Very good, Miss Jose.'

She turned to Meg. 'I want to hear what the piano sounds like, in case I have to play this afternoon.'

Pom! Ta-ta-ta, *tee*-ta! At the sound of the piano, Jose's face changed. She looked with eyes full of suffering at her mother and Laura as they came in. '*This life is weary*,' she sang.

'*A tear – a sigh.*
A love that changes,
And then – goodbye!'

But on the word 'goodbye', although the piano sounded desperately sad, a big, bright, completely unsympathetic smile appeared on Jose's face.

'Aren't I singing well today, Mummy?' she said happily, and started to sing again.

'This life is weary,
Hope comes to die.
A dream ...'

But Sadie had come in.

'What is it, Sadie?'

'Please, Miss Jose, cook says she needs the flags for the sandwiches.'

'The flags for the sandwiches, Sadie?' Mrs Sheridan said in a dreamy voice. And the children knew by her face that she hadn't got them. 'Let me see. Tell cook I'll get them to her in ten minutes.'

Sadie went.

'Now, Laura,' said her mother quickly, 'come with me into the smoking-room. I've got the names on the back of an envelope. You'll have to write them on the flags for me. Meg, go upstairs and brush your hair. Jose, go and dress immediately. Quickly, children, or I shall have to speak to your father about you. And Jose – if you go into the kitchen, try and calm cook down, will you? I'm quite frightened of her this morning.'

Mrs Sheridan found the envelope, and told Laura what to write on the flags for the sandwiches.

'Chicken and banana. Have you done that one?'

'Yes.'

'Egg and—' Mrs Sheridan held the envelope away from her. 'Fish. Can this possibly say fish?'

'Yes, mother dear,' said Laura, looking over her shoulder.

'Fish. It sounds absolutely horrible. Egg and fish.'

The flags were finished at last, and Laura took them to the kitchen. Jose was there, talking to the cook, who looked perfectly calm and cheerful.

'I have never seen such beautiful sandwiches,' Jose was saying enthusiastically. 'How many kinds are there?'

'Fifteen, Miss Jose.'

'Well, cook, I congratulate you.'

Cook smiled happily.

'Godber's has come,' said Sadie. She meant that the man from Godber's shop had brought the chocolate cakes. Godber's chocolate cakes were famous. Nobody ever made their own if they could buy Godber's.

'Bring them in and put them on the table, my girl,' ordered cook.

Sadie brought them in and went back to the door. Of course, Laura and Jose were far too grown up to care about chocolate cakes. All the same, those cakes looked nice. Very nice. Cook began arranging them on plates.

'Don't they remind you of all the parties we had when we were children?' said Laura.

'I suppose they do,' said Jose, who never liked to think about the past. 'They look delicious, I must say.'

'Have some, my dears,' said cook in her comfortable voice. 'Your ma won't know.'

Oh, impossible. Chocolate cake, and so soon after breakfast? It was unimaginable. All the same, two minutes later, Jose and Laura were licking chocolate off their fingers.

'Let's go into the garden,' suggested Laura. ' I want to see how the men are getting on with the marquee. They're such awfully nice men.'

But the door was blocked by cook, Sadie, Godber's man and Hans.

Something had happened.

Cook was making little 'tuk-tuk-tuk' sounds. Sadie had her hand over her mouth. Hans was trying so hard to understand that his eyes were closed tight. Only Godber's man was enjoying himself. It was his story.

'What's the matter? What's happened?'

'There's been a horrible accident,' cook said. 'A man's been killed.'

'A man killed! Where? How? When?'

But Godber's man wanted to tell the story.

'You know those little cottages just down the road from here, miss?' Of course she knew them. 'Well, there's a young fellow living there, Scott he's called, he's got a horse and cart. Something frightened the horse in town this morning, the cart turned over, and this fellow Scott was thrown out. He fell in the road on the back of his head. Killed.'

'Dead!' Laura stared at Godber's man.

'Dead when they picked him up,' Godber's man said with enjoyment. 'They were taking the body home as I was coming here.' Then he said, 'He's left a wife and five little ones.'

'Jose, come here!' Laura took her sister's hand and pulled her across the kitchen and through the door.

'Jose,' she said, 'how can we stop everything?'

'Stop everything, Laura!' cried Jose. 'What do you mean?'

'Stop the garden party, of course.' Why did Jose pretend not to understand?

But Jose was even more surprised. 'Stop the garden party? My dear Laura, don't be unreasonable. Of course we can't stop the party. Nobody expects us to. Don't be so silly.'

'But we can't possibly have a garden party with a man dead just outside the front gate.'

That really was silly, because the Sheridans' house was on a hill, and the cottages were right down at the bottom of the hill. There was a wide road between them. True, they were still much too near. They were not suitable neighbours for people like the Sheridans.

The cottages were ugly little low brown things. Nothing but rubbish grew in their gardens. Even the smoke coming from their chimneys looked poor and mean. The people who lived in them lived by washing other people's clothes, or mending shoes or cleaning chimneys. And they all had far too many children.

When the Sheridan children were little, they were not allowed to go near the cottages, in case they heard bad language or caught some awful disease. But now that they were grown up, Laura and Laurie sometimes walked past. It was dirty and unpleasant, but Laura and Laurie believed that they should experience all sides of life. They must go everywhere; they must see everything.

'Think how that poor woman will feel if she hears a band playing,' said Laura.

'Oh, Laura!' Jose began to be seriously annoyed. 'If you want to stop a band every time someone has an accident, you're going to have a very difficult life. I'm just as sorry about it as you are.' A hard look came into her eyes. She looked at her sister in the way she had looked when they were little girls fighting together. 'You won't bring a drunk workman back to life by stopping a party,' she said softly.

'Drunk! Who said he was drunk?' Laura said angrily. She said, just as she had done when she was little, 'I'm going to tell mother.'

'Please do, my dear,' said Jose sweetly.

'Mother, can I come into your room?' said Laura, standing with one hand on her mother's door.

'Of course, child. Why, what's the matter? You look quite pink.' Mrs Sheridan turned from her mirror. She was trying on a new hat.

'Mother, a man's been killed,' Laura began.

'*Not* in our garden?' said her mother.

'No, no!'

'Oh, how you frightened me!' Mrs Sheridan took off the big hat and smiled at her daughter.

'But listen, mother,' said Laura. Breathlessly, she told the awful story. 'Of course, we can't have our party, can we?' she said. 'The band, and everybody arriving. They'd hear us, mother; they're nearly neighbours!'

To Laura's great surprise, her mother behaved just like Jose. It was worse, because she seemed to be amused. She refused to take Laura seriously.

'But my dear child, be sensible. We only know about the accident by chance. If someone had died there normally – and I don't know how they keep alive in those dirty little holes – we'd still be having our party, wouldn't we?'

Laura had to agree, but she felt it was all wrong.

'Mother, isn't it really terribly heartless of us?' she asked.

'Darling!' Mrs Sheridan got up, holding the hat, and before Laura could stop her, she put it on Laura's head. 'My child,' she said, 'the hat is yours. It's much too young for me. You look

'Oh, how you frightened me!' Mrs Sheridan smiled at her daughter.

wonderful in it. Look at yourself!' And she held up a mirror.

'But mother,' Laura began again. She couldn't look at herself. She turned away from the mirror.

This time Mrs Sheridan became angry, just as Jose had done.

'You are being very stupid, Laura,' she said coldly. 'People like that won't expect us to cancel our party. And it's not very thoughtful of you to ruin the day for everyone else.'

'I don't understand,' said Laura, and she walked quickly out of the room and into her own bedroom. There, quite by chance, the first thing she saw was a lovely girl in the mirror, wearing a beautiful black and gold hat. She had never imagined that she could look like that.

Is mother right? she thought. And now she hoped that her mother was right. Am I being stupid? Perhaps it was stupid. For a moment she imagined that poor woman again, and the little children and the body being carried into the house. But now it seemed shadowy and unreal, like a picture in the newspaper. I'll remember it again after the party's over, she decided. And somehow that seemed to be the best plan …

Lunch was over by half-past one. By half-past two they were all ready to begin the party. The band had arrived, and sat in a corner of the tennis court.

'My dear!' screamed Kitty Maitland, 'aren't they all *too* like monkeys in their little red jackets!'

Laurie arrived from the office. When she saw him, Laura remembered the accident again. She wanted to tell him about it. If Laurie agreed with the others, it meant that they were right. And she followed him into the hall.

'Laurie!'

'Hallo!' He was half-way upstairs, but when he turned and saw Laura, he stopped and stared at her. 'My word, Laura! You look wonderful,' said Laurie. 'What an absolutely topping hat!'

Laura said quietly, 'Is it?' and smiled at Laurie. She didn't tell him about the accident.

Soon after that, people started to arrive. The band began to play; the hired waiters ran from the house to the marquee. Wherever you looked, there were couples walking, looking at the flowers, greeting friends, moving on over the lawn. They were like bright birds that had come to visit the Sheridans' garden for this one afternoon, on their way to – where? Ah, what happiness to be with people who are all happy, to shake hands, kiss, smile.

'Darling Laura, how well you look!'

'What a beautiful hat, child!'

'Laura, you look quite Spanish. I've never seen you look so lovely.'

And Laura, happy, answered softly, 'Have you had tea? Won't you have an ice-cream? The coffee and brandy ice-creams really are rather special.' She ran to her father and begged him: 'Daddy darling, can the band have something to drink?'

And the perfect afternoon slowly opened, slowly turned to the sun, and slowly closed like a flower.

'The most enjoyable garden party …'

'The greatest success …'

'Quite the most delicious …'

Laura helped her mother with the goodbyes. They stood side by side until all the guests had gone.

'All over, all over,' said Mrs Sheridan. 'Go and find all the others, Laura. Let's go and have some fresh coffee. I'm exhausted. Yes, it's been very successful. But oh, these parties, these parties! Why do you children insist on giving parties!' And they all sat down in the empty marquee.

'Have a sandwich, Daddy dear. I wrote the flag.'

'Thanks.' Mr Sheridan took a bite and the sandwich was

gone. He took another. 'Did you hear about a nasty accident that happened today?' he said.

'My dear,' said Mrs Sheridan, holding up her hand, 'we did. It nearly ruined the party. Laura wanted us to cancel everything.'

'Oh, mother!' Laura did not want them to laugh at her.

'It was a horrible thing, though,' said Mr Sheridan. 'The fellow was married, too. Lived in one of those cottages down there. Leaves a wife and a whole crowd of kids, they say.'

There was a long silence. Mrs Sheridan played with her cup. Really, it was most unfortunate that father had mentioned ...

Suddenly, she looked up. They still had all these perfectly good sandwiches and cakes which had not been eaten at the party. She had one of her clever ideas.

'I know,' she said. 'Let's send that poor creature some of this food. We'll pack a basket. All those children will love it. And I'm sure all the neighbours are calling in. How helpful it will be for her to have some extra food ready. Laura! Get me the big basket from the kitchen cupboard.'

'But mother, do you really think it's a good idea?' said Laura.

Again, how strange, she seemed to be different from them all. To take the left-over food from their party. Would the poor woman really like that?

'Of course! What's the matter with you today? An hour or so ago you were insisting on us being sympathetic.'

Oh well! Laura ran to get the basket. Her mother heaped food into it.

'Take it yourself, darling,' she said. 'Run down with it now. No, wait, take some lilies too. These lilies will seem like something really special to people of that kind.'

'She'll get her dress dirty if she takes flowers,' said Jose.

29

That was true. Just in time. 'Only the basket, then. And Laura!' – her mother followed her out of the marquee – 'whatever happens, don't ...'

'What, mother?'

No, it was better not to put ideas into the child's head. 'Nothing! Run along.'

It was beginning to get dark as Laura shut the garden gate. Below her, the road shone white. The little cottages were in deep shadow. How quiet it seemed after the excitement of the day. She was going down to a cottage where a man lay dead, and she couldn't believe it. Why couldn't she? She stopped for a moment. And it seemed that kisses, voices, laughter, the fresh smell of the lawn were somehow inside her. She had no room for anything else. How strange! She looked up at the pale sky, and all she thought was, 'Yes, it was the most successful party.'

She crossed the wide road. She was among the cottages. Men and women hurried past. Children played in the narrow streets. Noises came from inside the mean little houses. In some there was lamp-light, and shadows moved across the windows.

Laura bent her head and hurried on. She wished now that she had put on a coat. People were staring at her dress and her black and gold hat – oh, how she wished it was a different hat! It was a mistake to come here; she had known all the time that it would be a mistake. Should she go back, even now?

No, too late. There was the house. It must be this one. There were people standing outside. Beside the gate an old, old woman sat in a chair, watching. She had her feet on a newspaper. The voices stopped as Laura came near. They moved to one side to let her walk past. She felt that they were expecting her. They had known that she would come.

Laura felt very shy and frightened. 'Is this Mrs Scott's house?' she asked a woman, and the woman answered, with a strange little smile, 'It is, my girl.'

Oh, how she wanted to escape from this! She actually said out loud, 'Help me, God,' as she walked up the tiny path and knocked at the door. I'll just leave the basket and go, she decided. I won't even wait for them to empty it.

Then the door opened. A little woman in black appeared.

Laura said, 'Are you Mrs Scott?' But to her horror the woman answered, 'Come in, please, miss,' and she was shut in the passage.

'No,' said Laura, 'I don't want to come in. I only want to leave this basket. Mother sent—'

The little woman in the dark passage seemed not to hear her. 'This way, please, miss,' she said in an oily voice, and Laura followed her.

She found herself in a little low kitchen, lit by a smoky lamp. There was a woman sitting by the fire.

'Em,' said the little creature who had let her in. 'Em! It's a young lady.' She turned to Laura. 'I'm her sister, miss,' she said. 'You'll excuse her, won't you?'

'Oh, but of course!' said Laura. 'Please, please don't disturb her. I only want to leave—'

But at that moment the woman by the fire turned round. Her face – red-eyed and wet – looked terrible. She didn't seem to understand why Laura was there. What did it mean? Why was a stranger standing in the kitchen with a basket? And more tears fell from those poor red eyes.

'All right, my dear,' said the sister. 'I'll thank the young lady.' And she gave Laura an oily smile.

At that moment the woman by the fire turned round.

Laura only wanted to get out, to get away. She went out into the passage, a door opened, and she walked into the bedroom, where the dead man was lying.

'You'd like to see him, wouldn't you?' said Em's sister. 'Don't be afraid, my girl. He looks a picture. Not a mark on him. Come along, my dear.'

32

Laura went up to the bed.

A young man lay there, asleep – sleeping so deeply that he was far, far away from them both. So distant, so peaceful. He was dreaming. Never wake him up again. His eyes were closed, deep in his dream. What did garden parties and baskets and dresses mean to him? He was far away from all those things. He was wonderful, beautiful. While they were all laughing and the band was playing, this beautiful thing had come to the cottages. Happy ... happy ... All is well, said that sleeping face. This is what should happen. I am at rest.

But at the same time, it made you want to cry, and Laura couldn't go out of the room without saying something to him. She burst into tears, like a little girl.

'Forgive my hat,' she said.

And this time she didn't wait for Em's sister. She found her way out of the house, past all the people. At the corner of the street she met Laurie.

He appeared out of the shadows. 'Is that you, Laura?'

'Yes.'

'Mother was getting anxious. Was it all right?'

'Yes, quite. Oh, Laurie!' She ran to him and took his arm.

'I say, you're not crying, are you?' asked her brother.

Laura shook her head. She was.

Laurie put his arm round her shoulders. 'Don't cry,' he said, in his warm, loving voice. 'Was it awful?'

'No,' said Laura. 'It was absolutely wonderful. But Laurie—' She stopped. She looked at her brother.

'Isn't life,' she began, 'isn't life ...' But what life was, she couldn't explain. It didn't matter. She knew he understood.

'*Isn't* it, darling?' said Laurie.

Pictures

Eight o'clock in the morning. Miss Ada Moss lay in her narrow bed, staring up at the ceiling. Her room, which was right at the top of a tall house in Bloomsbury, smelled of wet clothes and face powder and the bag of fried potatoes she had brought in for supper the night before.

'Oh dear,' thought Miss Moss. 'I am cold. I wonder why I always wake up so cold in the mornings now. My knees and feet and my back – especially my back – are like ice. And I was always so warm in the old days. It isn't because I'm thin. I'm just as well-covered as I always was. No, it's because I don't have a good hot dinner in the evenings.'

She imagined a row of good hot dinners passing across the ceiling, each with a bottle of good strong beer.

'I'd like to get up now,' she thought, 'and have a big sensible breakfast.' Pictures of big sensible breakfasts followed the good hot dinners across the ceiling. Miss Moss pulled the blanket up over her head and closed her eyes. Suddenly, her landlady burst into the room.

'There's a letter for you, Miss Moss.'

'Oh,' said Miss Moss, in a voice which was much too friendly, 'thank you very much, Mrs Pine. It's very good of you to bring me my letters.'

'Oh, it's nothing,' said the landlady. 'I hope it's the letter that you've been waiting for.'

'Yes,' said Miss Moss brightly, 'yes, perhaps it is. I wouldn't be surprised.'

'Well, I'd be very surprised,' said the landlady. 'That's the truth. And can you open it right now, please. A lot of landladies wouldn't even ask – they'd just open it themselves. Things can't go on like this, Miss Moss, indeed they can't. First you tell me you've got the money to pay your rent, then you say you haven't, then there's a letter lost in the post or a theatre manager who's gone to Brighton but will be coming back soon – I'm sick and tired of it all, and I've had enough. At a time like this, too, with the price of everything sky-high and my poor boy away at the war in France! If you can't pay your rent, there's plenty of other people who would give me good money for a room like this. As my sister Eliza was saying to me only yesterday, Miss Moss, I've been much too soft-hearted with you!'

Miss Moss did not seem to be listening to this. She tore open the letter. It was from a film company.

'*No suitable parts for a lady of your experience at present,*' she read.

She stared at it for a long time before she spoke to her landlady.

'Well, Mrs Pine,' she said. 'I think you'll be sorry for what you've just said. This is from a theatre manager who wants to see me immediately about a part in a new musical show.'

But the landlady was too quick for her. She tore the letter out of Miss Moss's hand.

'Oh is it, is it indeed!' she cried.

'Give me back that letter. Give it back to me at once, you bad, wicked woman,' cried Miss Moss. She could not get out of bed because she had a hole in her nightdress.

'Well, Miss Moss,' said the landlady, 'if I don't get my money by eight o'clock tonight, you can get out of my house, *my lady*.'

The door banged and Miss Moss was alone. She threw back the bedclothes, and sat on the side of the bed, shaking with anger and staring at her fat white legs. 'The old cat,' she said, 'the rotten old cat!' Then she began to pull on her clothes.

'Oh, I wish I could pay that woman! Then I'd tell her what I think of her!' She suddenly saw her face in the mirror, and gave herself a little smile.

'Well, old girl,' she said, 'you're in trouble this time, and no mistake.' But the person in the mirror stopped smiling.

'You silly thing,' said Miss Moss. 'It's no good crying. You'll make your nose all red. Come on! Get dressed, and go out and find a job. That's what you've got to do.'

She picked up her bag and shook it. A few small coins fell out. 'I'll have a nice cup of tea at an ABC café before I go anywhere,' she decided. 'I've got enough money for that.'

Ten minutes later, a large lady in a blue dress and a black hat covered in purple flowers looked at herself in the mirror, and sang:

'Sweetheart, remember that hope never dies
And it al–ways is dark–est before sunrise.'

But the person in the mirror wouldn't smile at her, and Miss Moss went out.

When she came to the ABC café, the door was open. A man was carrying boxes of bread in, and two waitresses were combing their hair and talking.

'My young man came home from France last night,' one of the girls sang happily.

'Oh, I *say*! How topping for you!' cried the other.

'Yes, wasn't it! He brought me a sweet little brooch. Look, it's got "Dieppe" written on it.'

'Oh, I *say*! How topping for you!'

The man with the boxes of bread came in again, almost knocking Miss Moss over.

'Can I have a cup of tea, please?' she asked.

But the waitress went on combing her hair. 'Oh,' she sang, 'we're not *open* yet.' She turned to the other girl. 'Are we, dear?'

'Oh, *no*,' said the other waitress.

Miss Moss went out. 'I'll go to Charing Cross,' she decided. 'That's what I'll do. And I'll have coffee, not tea. Coffee's more filling. Those girls! Her young man came home; he brought her a brooch ...' She began to cross the road.

'Look out, Fatty!' shouted a taxi-driver. Miss Moss pretended not to hear.

'No, I won't go to Charing Cross,' she decided. 'I'll go straight to Kig and Kadgit. They open at nine. If I get there early, Mr Kadgit may have something for me ... "I'm so glad to see you, Miss Moss. I've just heard from a manager who wants a lady ... exactly the right part for you ... three pounds a week ... go and see him immediately. It's lucky you came so early."'

But there was nobody at Kig and Kadgit except an old woman washing the floor in the passage.

'Nobody here yet, Miss,' the old woman said.

'Oh, isn't Mr Kadgit here?' said Miss Moss. 'I'll sit down and wait for him, if I may.'

'You can't wait in the waiting-room, Miss. I haven't cleaned it yet. Mr Kadgit never comes in before eleven-thirty on a Saturday. Sometimes he doesn't come in at all.'

'How silly of me,' said Miss Moss. 'I forgot it was Saturday.'

'Mind your feet, please, Miss,' said the old woman. And Miss Moss was out in the street again.

The nice thing about Beit and Bithem was – it was always crowded. You walked into the waiting-room and you met everybody you knew. The early ones sat on chairs, and the later ones sat on the early ones' knees, while the men stood around the walls, talking and joking with the ladies.

'Hello,' said Miss Moss, in her friendly way. 'Here we are again!'

And young Mr Clayton did a couple of dance-steps and sang: 'Waiting for the Robert E. Lee!'

'Mr Bithem here yet?' asked Miss Moss, powdering her nose.

'Oh yes, dear!' cried all the girls together. 'He's been here for *ages*. We've been waiting for more than an *hour*!'

'Oh dear,' said Miss Moss. 'Any work for us, do you think?'

'Oh, a few jobs in Africa,' said young Mr Clayton. 'A hundred and fifty a week for two years, you know.'

'Oh!' cried the girls. 'Isn't he a *scream*? Isn't he *too* funny?'

A dark girl with a sad face touched Miss Moss on the arm. 'I just missed a lovely job yesterday,' she said. 'Six weeks on tour, and then the West End. The manager said I would have got it if I'd been a bit stronger-looking. He said the part was made for me – only I'm too thin.' She stared at Miss Moss, and the dirty, dark red rose on her hat looked as sad and disappointed as she was.

'Oh dear, that was awfully bad luck,' Miss Moss said, trying hard not to sound too interested. 'What was the show called, may I ask?'

But the sad, dark girl understood what Miss Moss wanted, and a mean look came into her heavy eyes.

'Oh, it wasn't a part for you, my dear,' she said. 'He wanted

38

someone young, you know, a dark Spanish type like me. I was too thin, that was the only problem.'

The door opened and Mr Bithem appeared. He kept one hand on the door, and held up the other for silence.

'Look here, ladies' – and here he paused and gave them his famous smile – 'and all you *boys*.' They all laughed loudly at that. 'I've got nothing for you this morning. Come back on Monday. I'm expecting several phone calls on Monday.'

Miss Moss pushed desperately through the crowd. 'Mr Bithem, I wonder if you've had any news from …'

'Now, let me see,' said Mr Bithem slowly, staring at her. He had seen Miss Moss four times a week for – how many weeks was it? 'Now, who are you?'

'Miss Ada Moss.'

'Oh yes, yes; of course, my dear. Not yet, my dear. Now I had a call for twenty-eight ladies today, but they had to be young and able to kick their legs up a bit. Come back the week after next – there'll be nothing before that.'

He gave her a big smile, all for herself, and touched her lightly on her fat arm before disappearing back into his office.

At the North-East Film Company they were waiting on the stairs. Miss Moss stood and waited next to a fair little baby-girl of about thirty, in a white hat with fruit all round it.

'What a crowd!' Miss Moss said. 'Is something special happening today?'

'Didn't you know, dear?' said the baby, opening her huge, pale eyes. 'There was a call at nine-thirty for pretty girls. We've all been waiting for hours. Have you worked for this company before?'

'No, I don't think I have,' said Miss Moss.

'They're a lovely company,' said the baby. 'A friend of mine has a friend who gets thirty pounds a day … Have you been in many films?'

'Well, I'm not really an actress,' said Miss Moss. 'I'm a trained singer. But things have been so bad lately that I've been doing a little acting.'

'It's *like* that, isn't it, dear?' said the baby.

'Have you been in many films?' said the baby.

'I had an excellent education at the College of Music,' said Miss Moss. 'I've often sung in West End shows. But I thought, for a change ...'

'Yes, it's *like* that, isn't it, dear?' said the baby.

At that moment a beautiful secretary appeared at the top of the stairs.

'Are you all waiting for news from the North-East Film Company?'

'Yes!' they all cried.

'Well, it's been cancelled. I've just had a phone call.'

'And I really *needed* that money,' a disappointed voice said.

The secretary had to laugh. 'Oh, there was no *money* in it,' she said. 'The North-East never pay their *crowd* people.'

There was only a little round window at the Bitter Orange Company. No waiting-room, nobody at all except a girl who came to the window and said, 'Well?'

'Can I see the manager, please?' Miss Moss said pleasantly.

The girl closed her eyes for a moment. Miss Moss smiled at her. The girl did not smile back. She frowned. She seemed to smell something unpleasant. Suddenly, she picked up a piece of paper and pushed it through the window at Miss Moss.

'Fill in this form!' she said, and banged the window shut.

'Can you ride a horse – drive a car – dive – fly a plane – shoot?' Miss Moss read. She walked along the street, asking herself those questions. A rough, cold wind was blowing. It pulled at her clothes, hit her in the back and then laughed cruelly in her face. It knew that she could not answer the questions.

In the Square Gardens, she found a rubbish basket, and dropped the form in it. Then she sat down on a bench and took out a little mirror to powder her nose. But the person in the mirror made an ugly face at her, and Miss Moss had to cry. She cried for a long time; it cheered her up wonderfully.

'Well, that's over,' she said. 'It's nice to be able to sit on this bench and rest my feet for a bit. And my nose will soon stop being red. Look at the birds! How close they come. I suppose someone feeds them. No, I've got nothing for you ...' She looked past them. What was that big building – the Café de Madrid? Oh, look at that poor child! Down he went with such a crash. Never mind! Up again! ... *If I don't get my money by eight o'clock tonight* ... Café de Madrid. 'I could just go in and sit there and have a coffee, that's all,' thought Miss Moss. 'Lots of artists go there, too. I might be lucky ... A dark handsome gentleman comes in with a friend, and sits at my table, perhaps ... "No, Julian, I've searched London for a singer who can take the part, and I just can't find the right person. You see, the music is difficult; have a look at it." ' And Miss Moss heard herself saying: 'Excuse me, but I happen to be a singer, and I have sung that part many times ... "Extraordinary! Come back to my studio and I'll try your voice now." ... Ten pounds a week ... Why should I feel frightened? It's not fear. Why shouldn't I go to the Café de Madrid? I'm an honest woman – I'm a professional singer. And I'm only trembling because I've had nothing to eat today ... "*You can get out of my house, my lady.*" ... Very well, Mrs Pine. Café de Madrid. They have music there in the evenings ... "Why don't they begin?" The singer has not arrived ... "Excuse me, I happen to be a singer; I have sung that music many times." '

It was almost dark in the café. Men, tall potted plants, red seats, white stone tables, waiters in black jackets. Miss Moss walked past them all and sat down.

Almost immediately, a very large gentleman wearing a very small hat came and sat opposite her.

'Good evening!' he said.

Miss Moss said, in her cheerful way: 'Good evening!'

'Fine evening,' said the large gentleman.

'Yes, very fine. Lovely, isn't it?' she said.

He waved a finger at a waiter. 'Bring me a large whisky.' Then he turned to Miss Moss. 'What's yours?'

'Well, I think I'll take a brandy, thank you very much.'

Five minutes later he turned to Miss Moss and blew a cloud of cigar smoke in her face.

'Like the hat,' he said, looking at the purple flowers.

Miss Moss blushed a deep pink and her heart began to beat very fast.

'I've always worn a lot of purple,' she said.

The large gentleman looked at her for a long time, tapping with his fingers on the table.

'I like a woman with a bit of meat on her bones,' he said.

Miss Moss, to her surprise, laughed quite loudly.

Five minutes later the large gentleman stood up.

'Well, am I coming to your place, or are you coming to mine?' he asked.

'I'll come with you, if you don't mind,' said Miss Moss, and she followed him out of the café.

The Little Governess

Oh dear, she wished she wasn't travelling at night. She would much rather have travelled by day, much rather. But the lady at the Governess Agency had said: 'Take an evening boat. Then you can get into a "Ladies Only" carriage on the train next day, and that will be much safer than sleeping in a foreign hotel. Don't leave your seat on the train except to go and wash your hands, and when you do that, *make sure* you lock the door. The train arrives at Munich at eight o'clock in the morning, and Frau Arnholdt says that the Hotel Grunewald is only one minute away. She will arrive at six the same evening, so you can have a nice quiet day to rest and practise your German. When you want something to eat, I advise you to go to the nearest baker's and get a cake and some coffee. You haven't been abroad before, have you?'

'No.'

'Well, I always tell my girls that it's better to believe that other people are all bad than all good. It sounds hard, but we've got to be women of the world, haven't we?'

It was nice on the boat. The attendant in the Ladies' Sitting Room was so kind, and changed her money for her and helped her to find a comfortable place to lie down. She lay there and watched the other passengers taking off their hats and arranging themselves for the night. The attendant put a green shade over the lamp and sat with her sewing on her knees. 'I like travelling

44

very much,' thought the little governess. She smiled as she fell asleep.

But when the boat stopped and she moved sleepily forward with all the people who knew where to go and what to do – then she felt afraid. Just a little afraid, just enough to wish – oh, that it was day and that she had someone, another woman, to travel with her.

'Tickets, please. Show your tickets. Have your tickets ready.'

She was off the boat. Immediately, a man in a leather hat came forward and touched her on the arm. 'Where for, Miss?' He spoke English – he must be a guard or an official of some kind with a hat like that. But then – without asking – he took her bag and started pushing past people, shouting: 'This way!'. He had such a rude, determined voice.

'But I don't want help with my bag.' What a horrible man! 'I don't want any help. I can carry it myself.' He walked so fast that she had to run, and she tried to pull the bag out of his hand. He took no notice, but continued down the long, dark platform and across the railway line. She was sure that he was a robber as she, too, crossed the railway line. On the other side – oh, thank goodness! – there was a train with 'Munich' written on it.

The man stopped. 'Here?' asked the rude voice.

'Yes, a Ladies' carriage.' She opened her little purse to find something small enough to give to this horrible man, while he threw the bag into an empty carriage. There was a 'Ladies Only' notice stuck to the window. She got into the train and gave him a coin.

'What's this?' shouted the man, looking at it angrily. 'What have you given me? It isn't enough!' Did he think that he could trick her like that just because she was a girl and travelling alone at night? Never, never! She kept her purse safe in her hand and refused, simply refused, to see or hear him.

'Ah no. Ah no. This is not enough. You make a mistake.' He jumped onto the train and threw the money back at her.

Trembling with terror, she put out an icy hand and took the money. 'That's all you're going to get,' she said. For a minute he stood, staring at her with his sharp eyes. Then he said something she did not understand, and disappeared into the dark. Oh, how thankful she was! How simply terrible that had been. As she stood up to see if her bag was all right, she saw her face in the mirror – white, with big, round eyes. 'You're all right now,' she said to the mirror-face, feeling that it was more frightened than she was.

People stood talking in groups on the platform; a strange light from the station lamps made their faces green. A boy was selling tea; a woman offered blankets for hire. White smoke floated in the dark air. 'How strange it all is,' thought the little governess, 'and the middle of the night, too.' She looked out from her safe corner of the carriage. She was not frightened any more, but proud that she had not given that man any money. 'I can take care of myself – of course I can. The great thing is not to—'

Suddenly there was a sound of men's voices and loud laughter. It came nearer. The little governess froze in her corner as four young men passed, staring in at her through the window. One of them, bursting with the joke, pointed at the 'Ladies Only' notice, and they all stopped to stare at the little girl in the corner. Oh dear, they were in the next carriage. She heard them talking and laughing, and then there was a sudden silence before one of them, a tall thin fellow with black hair, threw open her door.

'We invite you to share our carriage, Mademoiselle,' he said in French. She could see the others, standing behind him. She sat very straight and very still. 'Please be so kind,' said the tall man, and behind him one of the others exploded into screams of laughter. 'The young lady is too serious,' said the man. Then they all went laughing back to their carriage and she was alone again.

'Close the doors! Close the doors!' Someone ran up and down beside the train.

'I wish it wasn't dark. I wish there was another woman in my carriage. I'm frightened of the men next door.' The little governess looked out of the window, and saw – the man in the leather hat coming back again. His arms were full of luggage. But – what was he doing? He tore the 'Ladies Only' notice off the window, while an old man climbed into the carriage. 'But this is for ladies only.'

'Oh, no, Miss. You make a mistake.'

'Close the doors!' A whistle blew and the train started.

Tears came to her eyes, but through them she saw the old man taking off his hat. He looked very old. Ninety, at least. But he had a nice face, pink, with little blue eyes. And he asked her, so politely, 'Would you like me to move to another carriage, Mademoiselle?'

What, make him carry all those heavy things? She couldn't do that! 'No, it's quite all right.'

'Ah, a thousand thanks.'

The train left the station and rushed into the dark. She looked out of the window but could see nothing – just the occasional light on a hill or the shape of a tree. In the carriage next door the young men were singing – the same song again and again. 'I wouldn't dare go to sleep if I were alone in here,' thought the little governess. She was glad that the old man was there. Really, he looked so nice, sitting there, so straight-backed and neat, reading his German newspaper. Some old men were horrible, but he ... He put down his newspaper. 'Do you speak German, Mademoiselle?'

'Yes, a little,' said the little governess, blushing a deep pink.

'Ah, then, perhaps you would like to look at my papers. I have several here.'

First, she took off her hat, and put it safely with her bag. How kindly the old man watched her as her little hand turned over the big white pages. Her beautiful golden hair hung over her face. How sad to be a poor little governess and have such wonderful hair! Perhaps the kind old man was thinking that. Perhaps he thought, 'Poor little girl, travelling all alone at night. I wish I could be a grandfather to her and look after her!'

'Thank you very much,' she smiled, giving back the papers.

'But you speak German extremely well,' said the old man. 'You have been in Germany before, of course?'

'Oh, no, this is the first time I have ever been abroad at all.'

'Really! I am surprised. I had the feeling you had travelled a great deal before. Well, you will like Munich,' said the old man. 'Munich is a wonderful city. Museums, pictures, theatres, shops – you can find everything in Munich. I have travelled all over Europe, but I am always glad to get back to Munich.'

'I am not going to stay in Munich,' said the little governess shyly. 'I am going to be governess to a doctor's family in Augsburg.'

Ah, he knew Augsburg. A fine city, too. 'But you should take a little holiday in Munich before you go.'

'Oh, I am afraid I could not do *that*,' said the little governess seriously. 'Also, if one is alone ...' He understood. He, too, looked serious, and they were both silent after that. The train flew on through the night. It was warm in the carriage. There were distant noises of doors opening and closing, rain on the windows. She fell asleep.

A sudden loud noise woke her. What had happened? The old man sat frowning. 'Ha! ha! ha!' came from the next carriage.

'Such thoughtless young men,' said the old man. 'I am afraid that they have woken you up with their noisy behaviour.' No, not really. She wanted to wake up now. She looked at her watch. Half-past four. A cold blue light filled the window. She looked out on fields, white houses, trees. How pretty it was! How pretty and how different! Even those pink clouds in the sky looked foreign. She rubbed her cold hands together, and felt very happy.

The train began to go more slowly. It gave a long whistle. They were coming to a town. Tall pink and yellow houses went by. A woman threw open her window and stared out at the train. More women appeared. And then – Look! What lovely flowers – and at the railway station, too! Colours you would never see at home.

The train stopped. A fat woman walked along the platform, carrying baskets of strawberries. Oh, she was thirsty! She was very thirsty!

The old man got up, smiling, and put his coat on. She smiled back at him as he left the carriage. While he was gone, the little governess looked at herself in the mirror and arranged her hair like a girl who is quite old enough to travel by herself. But she was so thirsty! She opened the window, and the fat woman with the strawberries came to her immediately. 'How much?' asked the little governess. 'Oh, my goodness! Much too expensive!' And she sat down in her corner again.

A whistle blew. She hoped the old man would not be left behind. Oh, here he was! And she smiled at him like an old friend as he came back into the carriage, holding – a basket of strawberries! 'Mademoiselle, please accept these ... '

'What, are they for me?' She looked alarmed, uncertain.

'Certainly, for you,' said the old man. 'I myself am no longer able to eat strawberries. Please. Try one.'

'Oh, thank you!' she gasped. 'They look so delicious.'

'Eat them all up,' said the old man, looking pleased and friendly. They were so big and juicy that she had to take two bites of each one. The juice ran all over her fingers. While she was eating them, she pretended to herself that the old man was her grandfather. What a perfect grandfather he would be!

When she had finished the strawberries, she felt that she had known him for years. She told him all about Frau Arnholdt and the arrangements for meeting her in Munich. Frau Arnholdt would not arrive until the evening. He listened, and then he said, 'I wonder if you would let me show you a little of Munich today. Just the English Garden, and perhaps a museum ... It would be so much more pleasant for you than sitting in a hotel room, and it would give an old man a great deal of pleasure.'

She said 'yes' immediately, and only wondered later if that had been a sensible thing to do. After all, she really did not know him. But he was so old and so kind – not to mention the strawberries. And it was her last day, really, her last day to enjoy herself.

'I will take you to your hotel,' he said, 'and call for you there at ten o'clock.' He gave her a card with his name on it. So then

'Eat them all up,' said the old man.

everything was arranged, everything was all right. And the little governess began to feel excited at being abroad, and enjoyed looking out of the window at all the new and foreign things, and talking about them all to the kind old grandfather until they reached Munich. He guided her carefully through the crowds at the station, and took her straight to her hotel. 'I will call for you at ten o'clock,' he said, and then he was gone.

'This way, Miss,' said a waiter, who had been staring at the little governess and the old man.

She followed him up the stairs to a dark bedroom. Ugh! What an ugly, cold room! How horrible it would be to spend the whole day here! 'Is this the room Frau Arnholdt ordered for me?' asked the little governess.

The waiter kept staring at her – he seemed to think there was something *peculiar* about her. He began to whistle, then changed his mind. 'Certainly,' he said.

Well, why didn't he go? Why did he stare at her like that? 'Go away,' said the little governess, in her cold English way. His little eyes widened in surprise. 'Go away at once,' she repeated icily.

He went to the door, but then he turned round. 'And your gentleman friend,' he said. 'Shall I send him upstairs to you when he comes?'

❦

Big white clouds over the white streets – and sunshine everywhere. Tall trees on both sides of the streets, trams full of fat, smiling people, a sound of laughter from open windows. And beside her, looking neater and more beautifully brushed than ever, her grandfather, who was showing her Munich. She wanted to run, to hang on his arm, to cry, 'Oh, I am so very happy!'

He guided her across the roads, waiting while she looked at everything and watching her with his kind eyes. She ate bread and meat and drank beer out of a huge glass like a flower vase. It didn't make you drunk like English beer. And then they went to look at pictures in the museum.

When they came out of the museum, it was raining. The grandfather put up his umbrella, and held it over her as they walked to a restaurant to have lunch. 'It will be easier if you take my arm,' he said. 'And, you know, that is the custom here in Germany.' So she took his arm and walked beside him, and the walk was so interesting that he forgot to put the umbrella down even when the rain stopped.

After lunch they went to the English Garden. 'I wonder what the time is,' said the little governess. 'My watch has stopped. We've seen so many things that I feel it must be quite late.'

'Late!' he laughed, 'late! But there is so much more to see, and you have not yet tried our delicious ice-cream!'

'Oh,' cried the little governess, 'I have enjoyed myself more than I can say. It has been wonderful! But Frau Arnholdt is calling at the hotel for me at six, so I should be there by five.'

'And you shall be there, but first we will sit down in this café and have a chocolate ice-cream.'

She was happy again. The ice-cream slipped down beautifully, and she sat with her back to the clock that pointed to twenty-five minutes to seven. 'Really and truly,' she said, 'this has been the happiest day of my life.' Her grateful baby heart was full of love for her dear old grandfather.

When they left the English Garden, the day was almost over. 'You see those big buildings there,' said the old man. 'That is where I live – I and my old housekeeper who looks after me.' She was very interested. 'Now, before I take you back to your hotel, will you come up for a few minutes and see my little home?' Of course, she would love to.

The passage was quite dark. 'Ah, I suppose my old woman has gone out to buy me a chicken.' He opened the door, and shy but curious, she went into a strange room. She did not know quite what to say. It wasn't pretty, but it was neat, and, she supposed, comfortable for such an old man. 'Well, what do you think of my little home?' He took a bottle and two pink glasses out of a cupboard. 'If ever you want to spend one or two days in Munich, there will always be a place for you here, and an old man ready to look after you.' He poured some wine into the pink glasses, and his hand shook a little as he poured. It was very quiet in the room.

She said, 'I think I ought to go now.'

'But you will have a little glass of wine with me – just one tiny glass before you go?' said the old man.

'No, really, no. I never drink wine, or anything like that.' And although she was afraid she was being awfully rude, she was quite determined. 'No, really, please.'

'Well, will you sit here by me for five minutes while I drink your health?'

The little governess sat down on the edge of the sofa, and he sat beside her and drank. 'Have you really been happy today?' asked the old man, and he sat so close to her that she could feel his knee against hers. Before she could answer, he took her hands in his. 'And are you going to give me one little kiss before you go?' he asked, pulling her towards him.

It was a dream. It wasn't true! It wasn't the same old man at all. Ah, how horrible! The little governess stared at him in terror. 'No, no no!' she gasped, pulling away from him.

'One little kiss. A kiss. Just a little kiss, my dear.' He pushed his face into hers, his lips smiling. How his little blue eyes shone!

'Never – never. How can you!' She jumped up, but he was too quick for her, and he pushed her against the wall and pushed his hard old body against hers. Although she fought him, shaking her head desperately from side to side, he kissed her on the mouth. On the mouth! Where nobody had ever kissed her before …

She ran, ran down the street until she found a wide road with trams and a policeman standing in the middle. 'I want to get a tram to the station,' cried the little governess.

'Excuse me?'

'The station!'

'There – there's a tram now,' and he watched, very much surprised, as the little girl with her hat on one side and tears pouring down her face, jumped on to the tram and sat there with her hands over her mouth.

When the little governess reached the Hotel Grunewald, the same waiter who had shown her to her room was there, putting glasses on the tables. He seemed oddly pleased to see her and to answer her questions. 'Oh, yes, the lady came. I told her you had arrived and then gone out again with a gentleman. She asked me when you were coming back again – but of course I could not say. And then she went to see the manager.' He picked up a glass and examined it closely. He smiled as he put it down again.

'Where is the lady now?' asked the little governess, shaking so violently that she had to hold her handkerchief up to her mouth.

'How should I know?' cried the waiter, and he pushed past her to greet some new guests coming through the door of the hotel. 'That's it! That's it!' he thought. 'That will teach her.' And as he picked up the new guests' luggage, he repeated again the little governess's words, '*Go away. Go away at once*. Shall I! Shall I!' he shouted to himself.

Her First Ball

Leila found it hard to say exactly when the ball began. Perhaps it began in the car taking her there. It did not matter that she shared the car with the Sheridan sisters and their brother. She sat back in her own little corner of it, and away she went, past dancing houses and fences and trees.

'Have you really never been to a ball before, Leila? But how strange—' cried the Sheridan girls.

'We lived so far from anyone else,' Leila said softly. 'In the country we had no near neighbours.'

Oh dear, how hard it was to be calm like the others! She tried not to smile too much; she tried not to care. But everything was so new and exciting. Meg's roses, Jose's necklace, Laura's little dark head above her white dress – she would remember these things for ever.

Her cousin Laurie reached over and touched Laura on the knee.

'Listen,' he said. 'We'll do the third and ninth dances together, as usual. OK, darling?'

Oh, how wonderful to have a brother! Leila was so excited that she suddenly wanted to cry, because she was an only child, and no brother had ever said 'OK, darling?' to her; no sister would ever say, as Meg said to Jose at that moment, 'I've never seen your hair look so lovely as it does tonight!'

But there was no time to cry. They were at the hall already.

The street was bright with moving lights and happy faces; little white shoes chased each other like birds.

'Hold on to me, Leila; you'll get lost,' said Laura.

'Come on, girls, let's go straight in,' said Laurie.

Leila put her hand on Laura's arm, and somehow the crowd carried them along and pushed them past the big golden lamp, along the passage and into the little room marked 'Ladies'. Here it was even more crowded and noisy. Everyone was pushing forwards, trying to get to the mirror.

There was a big gas light in the ladies' room. It wouldn't wait; it was dancing already. When the door opened, it jumped up almost as high as the ceiling.

Dark girls, fair girls were combing their hair, opening and closing bags, fastening buttons. And because they were all laughing, it seemed to Leila that they were all lovely.

'Aren't there any hairpins?' cried a voice. 'I need some hairpins.'

'Be a darling and put some powder on my back,' cried someone else.

'But I *must* have a needle and cotton! I've torn miles off my skirt!' screamed a third.

Then a voice said, 'Pass them along, pass them along!' and the basket of dance programmes went from hand to hand. Lovely little pink and silver programmes with tiny pink pencils. Leila's fingers shook as she took one. She wanted to ask, 'Should I take one, too?' But then Meg cried, 'Ready, Leila?' and they pushed through the crowd towards the doors of the dance-hall.

The band was silent, waiting to begin playing, but the room was full of the noise of talking and laughter. Leila felt that even the little coloured flags which hung from the ceiling were talking.

She forgot to be shy. She forgot how, earlier that day, she had sat on her bed with one shoe off and one shoe on and begged her mother to ring up her cousins and say that she couldn't come. The feeling she had had, that she wanted to go home, to be back at her dark lonely house out in the country, suddenly changed to a feeling of complete happiness that she was here at this ball.

She looked at the shining, golden floor, the flowers, the coloured lights, and at the stage, with its red carpet and golden chairs, and the band ready to play, and she thought, 'How lovely! How simply lovely!'

All the girls stood together on one side of the doors, and the men stood on the other side. Older ladies, in dark dresses, walked with little careful steps over the shiny floor towards the stage.

'This is my little country cousin Leila. Be nice to her. Find her partners. I'm looking after her,' Meg was telling all the girls.

Strange faces smiled sweetly at Leila. Strange voices answered, 'Of course, my dear.' But Leila knew that the girls didn't really see her. They were looking at the men. Why didn't the men begin? What were they waiting for? They stood there, not talking, just smiling to themselves. Then quite suddenly, they were coming towards the girls, flying towards them over the golden floor.

A tall, fair man flew up to Meg, took her programme and wrote something in it. Meg passed him on to Leila. 'May I have the pleasure?' He wrote in her programme, smiled at her and moved on. Then a dark man came up to Leila, then cousin Laurie and a friend. Then quite an old man – fat and rather bald, too – took her programme and said, 'Let me see, let me see!' He looked at his programme, which was black with names, and at Leila's programme. He seemed to have so much trouble finding a free dance for her that Leila felt ashamed.

'Oh, please don't bother!' she said eagerly.

But the fat man wrote something in her programme and

'How lovely!' Leila thought. 'How simply lovely!'

looked at her again. 'Do I remember this bright little face?' he said softly. 'Have I seen this little face before?'

At that moment the band began playing; the fat man disappeared. He was carried away on a wave of music that flew over the shining floor, breaking the groups of people into couples and throwing them out to the corners of the room.

Leila had learnt to dance at school. Every Saturday afternoon, the girls were taken to a little hall where Miss Eccles (of London) gave her 'top quality' lessons. But the difference between that poor little hall – with an old woman banging on the piano and Miss Eccles shouting at the girls to lift their feet – the difference between that place and this wonderful place of music and golden light was so great that Leila felt she would die if she didn't dance soon.

'Our dance, I think.' Someone smiled and gave her his hand. She didn't need to die, after all. She floated away like a flower on a stream.

'Quite a good floor, isn't it?' said a voice close to her ear.

'I think it's most beautifully slippery,' said Leila.

'Excuse me?' The voice sounded surprised. Leila said it again. There was a tiny pause before the voice said, 'Oh, quite,' and they danced on.

He danced so beautifully. That was the great difference between dancing with men and dancing with girls, Leila decided. Girls bumped into you and stepped on your feet.

The flowers were no longer flowers; they were pink and white flags flying by.

'Were you at the Bells' last week?' the voice said. It was a tired voice. Leila wondered whether she should ask him if he needed to stop and rest.

'No, this is my first ball,' she said.

He gave a little laugh. 'Oh, I say!'

'Yes, it really is the first ball that I've ever been to.' Leila felt quite excited, just talking about it. 'You see, I've lived in the country all my life …'

At that moment the music stopped, and they went to sit down. Leila's partner did not say very much. He stretched out his legs, played with a buttton on his jacket and looked around the room. But it didn't matter. The band began to play again, and her second partner seemed to appear from nowhere.

'Floor's not bad,' said the new voice. 'Were you at the Neaves' last Tuesday?'

Again, Leila explained that this was her first ball. It was strange that her partners did not seem to find this more interesting. It was so exciting! Her first ball! She was at the beginning of everything.

'How about an ice-cream?' said her partner. And they went through the doors, down the passage to the supper-room. Leila's face felt hot, and she was terribly thirsty. How sweet the ice-creams looked on their little glass plates, and how deliciously cold the spoons were!

When they came back to the hall, the fat man was waiting for her by the door. It gave Leila quite a shock to see how old he was; he ought to be on the stage with the mothers and fathers. And when she compared him with her other partners, his clothes looked old, too, and not terribly clean.

'Come along, little lady,' said the fat man. He held her loosely, and they moved so slowly that it was more like walking than dancing. But he said nothing at all about the floor. 'Your first ball, isn't it?' he said.

'How *did* you know?'

'Ah,' said the fat man, 'that's what it is to be old! You see, I've been doing this for the last thirty years.'

'Thirty years!' cried Leila. Twelve years before she was born!

'Terrible to think about, isn't it?' the fat man said sadly. Leila looked at his bald head, and she felt quite sorry for him.

'I think it's wonderful that you can still dance so well,' she said kindly.

'Kind little lady,' the fat man said. He held her a little closer. 'Of course,' he said, '*you* won't be able to go on as long as this. Oh, no,' said the fat man, 'long before you're as old as I am, you'll be sitting up there on the stage in your nice black dress, watching. And these pretty arms will be little short fat ones.' The fat man shook his head sadly at the thought. 'And you'll smile just like those poor old dears up there, and point to your daughter, and tell the old lady next to you how some terrible man tried to kiss her at a ball. And your poor heart will ache, ache' – he held her even closer, to show how sorry he felt for that poor heart – 'because no one wants to kiss you now. And you'll say how awful these slippery floors are, how dangerous to walk on. Yes, little lady?' the fat man said softly.

Leila gave a light little laugh, but she did not feel like laughing. Was it – could it all be true? Was this first ball only the beginning of her last ball? The music seemed to change. It sounded sad, sad. Oh, how quickly things changed! Why didn't happiness last for ever? For ever wouldn't be a bit too long!

'I want to stop,' she said in a breathless voice. The fat man led her to the door.

'No,' she said, 'I don't want to go outside.'

She stood there by the wall, trying to smile. But deep inside her a little girl threw herself down on her bed and burst into tears. Why did he have to ruin it all?

'I say, you know,' said the fat man, 'you mustn't take me seriously, little lady.'

'Of course I don't!' said Leila, biting her lip.

More people stood up to dance. The band was getting ready to play again. But Leila didn't want to dance any more. She wanted to be at home, looking out of her bedroom window at the stars.

But then the lovely music started, and a young man came to dance with her. She decided to dance with him and then go, as soon as she could find Meg. Very stiffly, she walked out onto the dance-floor. But in a moment her feet simply danced away with her. The lights, the flowers, the dresses, the pink faces, all became one beautiful flying wheel. When her next partner bumped her into the fat man, she just smiled at him happily. She didn't even recognize him again.

The Woman at the Store

All that day the heat was terrible. The wind blew close to the ground, lifting the white dust from the road and driving it into our faces. The horses moved wearily, half-blinded by the dust. The pack-horse was sick – with a big sore place on her back. She kept stopping, looking too tired to go on. Hundreds of birds screamed high overhead. There was nothing to see except mile after mile of coarse grass, with the occasional purple flower or grey-green bush.

Jo rode ahead. He wore a blue shirt and a white handkerchief round his neck, with a red pattern that looked like bloodstains on it. For once, he was not singing,

> 'I don't care, for don't you see,
> My wife's mother was in front of me!'

He had sung it every day for a month; now we felt lost without it.

Jim rode beside me, white-faced. He kept licking his dry lips. We had not spoken much since dawn.

'My stomach needs some real food inside it,' said Jo. 'Now, Jim, where's this store you keep talking about? You say you know a fine store with a paddock for the horses, and a creek, and a friend of yours waiting there with a bottle of whisky to share. I'd like to see that place, I really would …'

Jim laughed. 'Don't forget, Jo, there's a woman, too, with

64

blue eyes and yellow hair, and something else to share with you. Don't forget that!'

'The heat's making you crazy,' said Jo. We rode on. I fell half asleep, and dreamed that I was back home with my mother. I woke up to find that we were arriving somewhere.

We were on a hill, and below us there was a building with an iron roof. It stood in a garden; there was a paddock, a creek and some trees. Smoke rose from the chimney, and as I looked, a woman came out, followed by a child and a yellow sheep dog.

The horses found a final burst of speed, and Jo began singing, *'I don't care, for don't you see …'*

The sun came through the clouds and shone on the woman's yellow hair and on the gun she was carrying. The child hid behind her, as we got off our horses, and the yellow dog ran into the building.

'Hallo!' screamed the woman. 'The kid said there were three brown things coming over the hill. I ran out quick, I can tell you, to see what it was.'

'Where's your old man?' asked Jim.

The woman looked away, frowning. 'Gone shearing. Been away a month. You going to stop here? There's a storm coming.'

'Of course we are,' said Jo. 'So you're on your own, are you?'

She stood, looking from one to the other of us, like a hungry bird. I smiled to myself at the way the men had joked about her. Certainly she had blue eyes and yellow hair, but she was so ugly! Her hands were rough and red, and her stick-like legs were pushed into a pair of dirty old boots.

'I'll put the horses in the paddock,' said Jim. 'Got any horse medicine? One of them's got a sore back.'

'Wait a second.' The woman breathed deeply. Then she

She looked from one to the other of us, like a hungry bird.

shouted violently, 'You can't stop here! You've got to go. I've got nothing for you!'

'God help us!' said Jo heavily. He pulled me to one side. 'Gone crazy,' he said. 'On her own too much, if you know what I mean. Show some sympathy, and she'll change her mind.'

But there was no need for sympathy. She changed her mind anyway. 'Stay if you like,' she said. Then she turned to me. 'Come on – I'll give you the medicine for the horse.'

We went up the garden path. The yellow dog lay across the door, and she kicked it out of the way.

'The place isn't tidy. Had no time. Been ironing. Come in.'

It was a large room. The walls were covered in pictures cut from magazines. There was a table, some broken chairs, a pile of clothes she had been ironing. A door led into the store; through another door I saw the bedroom.

She left me there and went into the store for the medicine. I could hear her talking to herself. 'Now where did I put that bottle?' Down in the paddock Jo was singing, while Jim put up the tent. The sun was going down. There are no long evenings in our New Zealand days; the sun goes down and half an hour later it's night.

Sitting alone in that ugly room, I felt afraid. The woman was a long time. What was she doing in there? 'What a life!' I thought. 'Imagine living here all alone with that child and that dog. Mad? Of course she's mad! I wonder how long she's been here – I wonder if she'll talk to me.'

'What was it you wanted?' she shouted from the store.

'Some medicine for the horse.'

'Oh, I forgot what I was looking for. I've got it now.'

She came out and gave me a bottle.

'My, you look tired, you do. Shall I make you a few scones for supper? I've got some meat you can have, too.'

'All right.' I smiled at her. 'Bring the kid down to the paddock and eat with us.'

'Oh, no,' she said, shaking her head. 'I'll send the kid down with the food and some milk. Want some scones to take with you tomorrow?'

'Thanks.'

She came and stood by the door.

'How old is the kid?' I asked.

'Six next Christmas. Had a lot of trouble with her. Always sick when she was a baby, she was.'

'She doesn't look much like you. Is she like her father?' I asked.

'No!' she shouted. 'She's like me. Any fool could see that!'

I went down to the paddock and gave Jim the medicine for the sick horse. Jo had washed. He was combing his wet hair, smiling to himself.

I went to the end of the paddock, past the trees, and washed in the creek. The water was clear and soft as oil. I lay in the water and looked up at the trees. The air smelled of rain.

When I got back to the tent, Jim was lying by the fire. I asked him where Jo was.

'Didn't you see how he'd cleaned himself up?' said Jim. 'He said to me before he went off to find her, "She isn't much, but she's a woman. She'll look good enough in the dark!"'

'You told us she was pretty,' I said. 'That wasn't exactly true!'

'No, listen,' Jim said. 'I don't understand what's happened to her. I haven't been here for four years. I used to know the husband well. A fine, big fellow. And she worked in bars on the

West Coast – she was as pretty as a doll. Told me once she knew a hundred and twenty-five different ways of kissing!'

'Oh, Jim, she can't be the same woman!'

'Of course she is. I can't understand it. I think the old man's gone off and left her. That's just a lie about shearing!'

Through the dark we saw the kid coming towards us with a basket of food and some milk. I took them from her.

'Come here,' Jim said to her.

She went to him. She was a tiny, thin kid, with white hair and weak, pale blue eyes.

'What do you do all day?' asked Jim.

She stuck one finger in her ear. 'Draw.'

'What do you draw? Leave your ear alone!'

'Pictures.'

'What of? Cows and sheep?'

'Everything. I'll draw you when you're gone, and your horses and the tent, and that one' – she pointed at me – 'with no clothes on in the creek. I saw her but she couldn't see me.'

'Thanks a lot. How nice of you,' said Jim. 'Where's your Dad?'

'I won't tell you,' the kid said. 'I don't like your face.' She stuck a finger in the other ear.

'Here,' I said. 'Take the basket and go and tell the other man that supper's ready.'

She ran off and we started eating. We had finished before Jo arrived. He was very red-faced and cheerful, and he had a whisky bottle in his hand.

'Have a drink, you two,' he shouted. 'She wants us all to go and drink with her tonight.' He waved one hand in the air. 'We're good friends, her and me.'

'I can believe that!' laughed Jim. 'But did she tell you where her old man's gone?'

Jo looked up. 'Shearing,' he said. 'You heard her, you fool.'

The woman had tidied the room. She had even put flowers on the table, beside the oil lamp, the glasses and the whisky bottle. The kid was drawing on a piece of wrapping paper.

The woman's hair hung loose. Her face was pink and her eyes shone. She sat with her feet touching Jo's under the table. In the hot room, with insects flying round the lamp, we all got slowly drunk.

The woman was shouting. 'Six years I've been here,' she told us, 'and it's broken me, living here. I told him, it's broken me, taken away everything I had. Left me with this kid and nothing else. Trouble is,' she went on, 'he left me alone too much. He'd go off for weeks, leave me all alone here. He'd never stay long.'

'Ma,' said the kid, 'I drew a picture of them on the hill, and you and me and the dog.'

'Shut your mouth!' shouted the woman.

Suddenly there was lightning, followed by the crash of thunder.

'Good thing the storm's come,' said Jo. 'I've been feeling it in the air for days.'

'Where's your old man now?' asked Jim slowly.

Her head dropped forward onto the table. 'He's gone shearing and left me all alone again,' she cried.

'Watch the glasses,' said Jo. 'Come on, have another drink. No use crying about it.'

She dried her eyes and took the glass. 'It's a lonely life for a woman,' she said. Jo took her hand.

Every minute the lightning grew brighter and the thunder sounded nearer. I got up and went over to the kid, who immediately hid her drawings by sitting on them. 'You're not to look,' she said.

'Oh, come on, show us.' Jim came over to us, and we were just drunk enough to joke and laugh the kid into showing us the pictures. They were extraordinary drawings for a child to do – clever, but very nasty. No doubt about it, the kid's mind was diseased. While we looked at the pictures, she got madly excited, laughing and trembling all over.

'Ma!' she screamed. 'Now I'm going to draw what you told me I must never draw – now I'm going to!'

The woman rushed at her and hit her on the head.

'You'll get worse than that if you dare say that again!' she shouted.

Jo was too drunk to notice, but Jim caught the woman by the arm. The kid did not make a sound.

We listened to the thunder. Then the rain began to fall, hitting the iron roof like bullets.

'You'd better sleep here, not in the tent,' said the woman.

'Good idea,' said Jo quickly.

'Go and get your things from the tent. You two can sleep in the store with the kid. Mr Jo can have this room.'

It sounded a crazy arrangement, but nobody said anything. Jim and I took a lantern and went down to the tent. We ran through the rain, laughing and shouting like two children who are having a wonderful adventure.

When we came back, the kid was already in the store, lying on a blanket. Jo shouted, 'Good night, all!' We took a lamp and closed the door of the store.

Jim and I sat down on two packing cases. We looked around at the bags of potatoes, the smoked meats hanging from the ceiling, the advertisements for coffee on the walls – and couldn't stop laughing. The kid sat up and stared at us. We took no notice of her.

'What are you laughing at?' she said uneasily.

'You!' shouted Jim. 'You, and this whole place, my child.'

She screamed with anger and beat herself with her hands. 'I won't be laughed at, I won't!'

'Go to sleep, Miss, or do some drawing,' said Jim. 'Look, here's a pencil and a bit of paper.'

Through the noise of the rain we heard Jo's footsteps in the next room, then the sound of a door opening and closing.

'It's a lonely life for a woman,' whispered Jim.

'A hundred and twenty-five different ways!'

The kid threw the piece of paper at me. 'There you are,' she said. 'I've done it because Ma shut me in here with you two. The thing she said I never ought to draw. I drew the one she said she'd shoot me if I did. I don't care! I don't care!'

The kid had drawn a picture of the woman shooting a man and then digging a hole to bury him in.

She threw herself to the floor, and rolled around, biting her fingers.

Jim and I sat until dawn with the drawing beside us. The rain stopped, and the little kid fell asleep, breathing loudly. We got up and went down to the paddock. A cold wind was blowing – the air smelled of wet grass. Just as we got on to the horses, Jo came out of the building – he waved to us to ride on.

'I'll catch you up later!' he shouted.

A bend in the road, and the whole place disappeared.

Millie

Millie stood and watched until the men disappeared from view. When they were far down the road, Willie Cox turned round on his horse and waved to her. But she didn't wave back. Not a bad young fellow, Willie Cox, but a bit too free and easy in his ways. Oh, my word! It was hot. Hot enough to fry your hair.

Millie put her hand up to keep the sun out of her eyes, and looked out over the dry, burnt paddocks. In the distance along the dusty road she could see the horses, like brown flies jumping up and down. It was half-past two in the afternoon. The sun hung in the pale blue sky like a burning mirror, and away beyond the paddocks the blue mountains trembled and jumped like the sea.

Sid wouldn't be back until half-past ten. He had ridden over to the town with four of the farm boys, to help find the young fellow who'd murdered Mr Williamson. Such a terrible thing! And Mrs Williamson left alone with all those kids. Strange! She couldn't believe that Mr Williamson was dead. He was such a joker. Always making people laugh.

Willie Cox said they'd found him in one of the farm buildings, shot bang through the head. The young English fellow who was with the Williamsons to learn about farming had disappeared. Strange! Why would anyone shoot Mr Williamson? He was so popular. My word! What would they do to that young man when they caught him? Well, you couldn't feel sorry for him. As

73

Sid said, if they didn't hang him, he could just go out and kill someone else. There was blood all over the place. Willie Cox said he got such a shock when he saw it, that he picked a cigarette up out of the blood and smoked it. My word! He must have been half crazy.

Millie went back into the kitchen. Slowly, she washed the dinner plates. Then she went into the bedroom, stared at herself in the piece of mirror, and dried her hot, wet face with a towel. What was the matter with her that afternoon? She wanted to cry – about nothing! She decided to change her clothes and have a good cup of tea. Yes, that would help.

She sat on the side of the bed and stared at the coloured picture on the wall, *Garden Party at Windsor Castle*. In the middle of green lawns and shady trees sat Queen Victoria, with ladies in flowery dresses all around her. Behind them you could see the castle, with British flags flying from its towers. 'I wonder if it really looked like that.' Millie stared at the flowery ladies, who smiled coolly back at her. 'I wouldn't want their lives. Running round all day after the old Queen ... '

On the table that Sid had made for her from packing cases, there was a photograph of her and Sid on their wedding day. Now that was a nice picture! She was sitting in a chair in her white dress, with Sid standing with one hand on her shoulder, looking at her flowers. Behind them there was a waterfall, and Mount Cook in the distance, covered with snow. She had almost forgotten her wedding day. Time passed so quickly, and with nobody to talk to ... 'I wonder why we never had kids ... Well, *I've* never missed them. Perhaps Sid has, though. He's softer than me.'

Then she sat quiet, thinking of nothing at all, with her red hands on her knees. *Tick-tick* went the clock in the silent kitchen. Quite suddenly, Millie felt frightened. A strange trembling started inside her – in her stomach – and then spread all over to her knees and hands. 'There's somebody outside.'

She went softly into the kitchen. Nobody there. The back door was closed. She stopped and listened, and the furniture seemed to stretch and breathe … and listen, too. There it was again – something moving, outside. 'Go and see what it is, Millie Evans.'

She ran to the back door, opened it, and just saw somebody run and hide behind the wood pile. 'Who's there?' she called in a loud, brave voice. 'Come out! I seen you! I know who you are. I've got my gun.' She was not frightened any more. She was terribly angry. Her heart banged like a drum. 'I'll teach you to frighten a woman,' she shouted, and she took a gun and ran out of the house, over to the wood pile.

A young man lay there, on his stomach, with one arm across his face. 'Get up!' She kicked him in the shoulders. He didn't move. 'Oh, my God, I believe he's dead.' She knelt down and rolled him onto his back. She sat in the dust, staring at him; her lips trembled with horror.

He was not much more than a boy, with fair hair and a light beard on his chin. His eyes were closed, his face covered in dirt and dust. He wore a cotton shirt and trousers; there was blood on one of his trouser-legs.

'I *can't*,' said Millie, and then, 'You've got to.' She bent over and felt his heart. 'Wait a minute,' she whispered, 'wait a minute,' and she ran into the house for brandy and a bucket of water. 'What are you going to do, Millie Evans? Oh, I don't know. I never saw anyone unconscious before.' She knelt down,

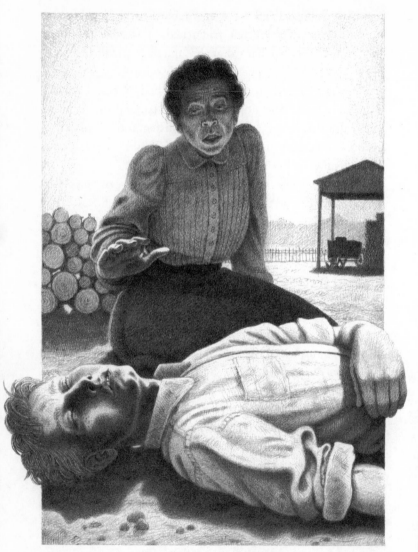

She sat in the dust, staring at him.

put her arm under the boy's head, and poured brandy between his lips. It ran out at the sides of his mouth. She took a cloth and washed his face and neck with the cool water. Under the dirt and dust, his face was as white as the cloth, thin, and marked by little lines.

A strange and terrible feeling took hold of Millie Evans. Deep inside her chest, it grew like a plant after rain, and burst painfully into leaf. 'Feeling better? All right, are you?' The boy breathed sharply, his eyes opened, and he moved his head from side to side. Millie touched his hair. 'Feeling fine now, aren't you?' The pain in her chest made her breathless. 'It's no good crying, Millie Evans. You've got to be sensible.' Suddenly he sat up and pulled away from her, staring at the ground. 'There, there,' cried Millie, in a strange, shaky voice.

The boy turned and looked at her, still not speaking. His eyes were so full of pain and terror that she had to shut her teeth together hard to stop herself crying. After a long pause he said, in the voice of a little child talking in his sleep, 'I'm hungry.' His lips trembled.

She stood up. 'Come on into the house and have a proper meal,' she said. 'Can you walk?'

'Yes,' he whispered, and followed her slowly to the door. Then he stopped. 'I'm not coming in,' he said. He sat down in the shade of the house.

Millie watched him. 'When did you last eat?' He shook his head. She went and put meat and bread and butter on a plate, but when she brought it to him, he was standing up, looking around. He did not take the plate of food she held out to him. 'When are they coming back?' he whispered.

At that moment she knew who he was. She stood there,

holding the plate, staring. He was Harrison, the English fellow who'd killed Mr Williamson. 'I know who you are,' she said, very slowly. 'I must have been blind not to see it from the start.'

He made a movement with his hands, which seemed to say, 'That's all nothing.' Again, he asked, 'When are they coming back?'

And she meant to say, 'Any minute now. They're on their way now.' Instead, she said to the poor frightened face, 'Not until half-past ten.'

He sat down and closed his eyes. Tears ran down his face. Just a kid. And all those men after him. 'Try a bit of meat,' Millie said. 'It's what you need. Get some good food in your stomach.' She sat down beside him, with the plate of food on her knees. 'Here – try a bit.' She broke the bread and butter into little pieces, and she thought, 'They won't catch him. Not if I can stop them. Men are all rotten. I don't care what he's done or not done. Do what you can to help him, Millie Evans. He's only a sick kid.'

Millie lay on her back in bed, with her eyes open, listening. Sid turned over, pulled the sheet round him and said, 'Good night, old girl.' She heard Willie Cox and the other fellows drop their clothes on the kitchen floor, and then their voices, and Willie Cox saying, 'Lie down, lie down, you little devil,' to his dog.

The house grew quiet. She lay there and listened. It was hot. She was frightened to move, because of Sid. 'He must escape, he must. I don't care about the law and all that rubbish they've been talking about,' she thought angrily. She listened to the silence. He ought to be moving ...

Before there was any sound from outside, Willie Cox's dog got up and went to the back door. A feeling of terror rose in

Millie. 'What's that dog doing? What a fool that young fellow is with a dog here. Why doesn't he lie down and sleep?' The dog stopped, but she knew it was listening.

Suddenly, with a sound that made her cry out in horror, the dog started barking and rushing about. 'What's that? What's happening?' Sid got out of bed.

'It's nothing, it's only Willie's dog. Sid, Sid!' She took his arm, but he pushed her away.

'By God, there's something out there!' Sid quickly pulled his trousers on. Willie Cox opened the back door, and the dog rushed madly out of the house.

'Sid, there's someone in the paddock,' one of the men shouted.

'What is it — what's that?' said Sid. 'Here Millie, take the lantern. Willie! There's someone in with the horses!'

The men ran out of the house, and at the same moment, Millie saw Harrison rush across the paddock on Sid's horse and down the road.

'Millie, bring that lantern, quick!' She ran out in her night-dress to give it to him. They were away down the road in a second.

And as she watched Harrison in the distance, and the men rushing after him, a strange and crazy delight came to her, drowning all other feelings. She ran into the road – she laughed and screamed and danced in the dust, waving the lantern in the air.

'After him, after him, Sid! Catch him, Willie! Go on, go on! Shoot him down! Shoot him!'

The Lady's Maid

Eleven o'clock. A knock at the door.

... I hope I haven't disturbed you, madam. You weren't asleep – were you? But I've just given my lady her tea, and there was a nice cup left, so I thought perhaps ...

... Not at all, madam. I always make a cup of tea at this time. She drinks it in bed after she's said her prayers, to warm her up. I start to boil the water when she kneels down, and I always say to the water, 'Now, don't be ready too quick.' But it always boils long before my lady finishes her prayers. You see, madam, we know such a lot of people, and my lady has to mention all their names in her prayers – every one. She keeps all their names in a little red book. Oh dear! Whenever we have a new visitor, and my lady says after they've gone, 'Ellen, bring me my little red book,' I feel quite wild, I really do.

And do you know, madam, she kneels right down on the hard carpet. It makes me worried sick to see it, knowing her the way I do. I've tried putting a soft woollen blanket down. But the first time I did it she gave me such a look – such a sweet, suffering look it was, madam. 'Did our Lord Jesus have soft woollen blankets, Ellen?' she said. But – I was younger then – I wanted to say, 'No, but our Lord Jesus wasn't as old as you, and He didn't have your poor bad back!' Terrible of me, wasn't it? But she's *too* good, you know, madam. When I went in to see if she was all right just now, and saw her lying there asleep – so pretty – I couldn't

80

help thinking, 'Now you look just like your dear mother on her deathbed.'

... Yes, madam, I took care of everything after the death. Oh, she did look sweet. I did her hair – ever so soft and pretty it was – and I put the most lovely flowers all round her head. She looked a picture! I shall never forget those flowers.

... Only the last year, madam. She came to live with us after she became a bit – well – forgetful, as they say. Of course, she was never dangerous; she was the sweetest old lady. But what happened was – she thought she'd lost something. She couldn't sit still. All day long she used to search the house, up and down, up and down. When she saw me, she'd say, 'I've lost it; I've lost it,' just like a child. And I'd say, 'Come along and we'll have a nice game of cards.' But she'd take my hand – I was a favourite of hers – and whisper, 'Find it for me, Ellen. Find it for me.' Sad, wasn't it?

... No, she never got any better, madam. The last thing she ever said was – very slow, 'Look in – the – Look – in –' And then she was gone.

... No, madam, I never noticed it. Perhaps some girls. But you see, it's like this. I've got nobody but my lady. My mother died when I was four, and I lived with my grandfather, who had a hairdresser's shop. I used to sit under a table in the shop, brushing my doll's hair. I suppose I was copying the assistants. They were really kind to me. I used to sit there all day, as quiet as can be – the customers never knew I was there.

... But one day I managed to get a pair of scissors, and – would you believe it, madam? – I cut off all my hair! What a little monkey! Grandfather went mad, he was so angry. He picked up the red-hot curling tongs – I shall never forget it – caught hold of my hand and shut my fingers in the tongs. 'That'll teach you!' he said. It was an awful burn. You can still see the mark.

... Well, you see, madam, he'd been so proud of my hair. He used to sit me in a high chair, before the customers came, and give me such a beautiful hair-do. I remember the assistants all standing and watching. Grandfather used to give me a penny to sit still for him. But he always took it back afterwards. Poor Grandfather!

When he burned my hand, I was so frightened! Do you know what I did, madam? I ran away. Yes, I did, right down the street and round the corner. Oh dear, I must have looked funny, with my hand rolled up in my skirt and my hair sticking out all over my head. People must have laughed when they saw me ...

... No, madam, Grandfather never forgave me. He couldn't even eat his dinner if I was in the room. So my aunt gave me a home. She was a dress-maker. Tiny little woman, she was. She had to stand on a chair to measure some of her ladies. And it was when I was helping her that I met my lady ...

... Not so very young, madam. I was already thirteen. And I don't remember ever feeling I was – well – a child. You see, there was my uniform, and one thing and another. My lady insisted on my wearing a proper uniform from the start.

... Oh yes – once I felt like a child! That was – funny! It was like this. My lady had her two little nieces staying with her, and there was a fair in a park nearby. 'Now, Ellen,' she said, 'I want you to take the two young ladies for a ride on the donkeys.'

Off we went. Quiet little girls, they were. They both held my hand. But when we came to the donkeys, they were too shy to go for a ride. So we stood and watched. Those donkeys were so beautiful! They were a lovely silver-grey colour and they had little red saddles and bells on their ears. And quite big girls – older than me – were riding them. And they looked so nice, just enjoying themselves. I don't know why, but when I saw those donkeys, with their little feet and their eyes – so gentle – and their big soft ears – well, I wanted more than anything in the world to ride on a donkey's back!

… Of course, I couldn't. I had to look after my young ladies. But all the rest of the day I thought about those donkeys. I had nothing but donkeys on my mind. I felt I would burst if I didn't tell someone; and there was no one to tell. But when I went to bed – I was sleeping in cook's bedroom at the time – as soon as the lights were out, I could see them again. My donkeys, with their neat little feet and sad eyes.

Well, madam, would you believe it, I waited for a long time and pretended to be asleep, and then I suddenly sat up and said, as loud as I could, '*I want to ride on a donkey. I want a donkey-ride!*' You see, I had to say it, so I pretended I was talking in my sleep. That's just what a silly child would do, isn't it?

… No, madam, never now. Of course, I wanted to when I was younger. But I never did. He had a little flower-shop. Funny, wasn't it? I've always loved flowers. We were having a lot of parties in the house at that time, and I was in and out of the flower-shop all the time. And Harry and I (his name was Harry) started arguing about what flowers were best – and that was how it began.

Flowers! You wouldn't believe it, madam, the flowers he used to give me. It was lilies more than once, and that's the honest truth. Well, of course, we were going to get married and live in rooms over the shop. I was going to arrange the flowers in the shop-window. Oh, how often I've arranged that window! Not really, of course, madam. Just in my dreams. I've done it all red and green for Christmas, and with a lovely star for Easter all made out of daffodils. I've put – well, that's enough about that.

The day came when we were going to choose our furniture. Shall I ever forget it? It was a Tuesday. My lady wasn't very well that afternoon. She didn't say anything, of course – she never complains. But she kept asking me if it was cold, and rubbing her little hands together. I knew she wasn't well. I didn't want to leave her, and I said, 'Shall I tell him that we'll go another day?'

'Oh, no, Ellen,' she said, 'you can't disappoint your young man.' So sweet, madam, never thinking about herself. It made me feel worse than ever.

And then she dropped a little perfume bottle, madam, and she tried to bend right down and pick it up herself – a thing she never did. 'Whatever are you doing!' I cried, and I ran to stop her.

'Well,' she said, smiling, 'I shall have to get used to doing things for myself now.' Oh, madam, I almost burst into tears when she said that! I couldn't stop myself, and I asked her if she would rather I didn't get married.

'No, Ellen,' she said – that was how she spoke, madam, just like that – 'No, Ellen, not for the world!' But while she said it, madam, I was looking in her mirror. Of course, she didn't know I could see her, and she put her little hand on her heart just like her dear mother used to, and she looked so sad ... Oh, *madam*!

When Harry came, I had his letters all ready, and the ring, and a dear little silver brooch with a heart on it that he'd given me. I opened the door to him. I never gave him time to say a word. 'There you are,' I said. 'Take them all back,' I said, 'it's all over. I'm not going to marry you,' I said, 'I can't leave my lady.' White! He turned as white as a woman. I had to shut the door, and I

She put her little hand on her heart, and she looked so sad.

stood there, shaking all over, until he went. Then I opened the door and I ran out into the middle of the road, and I just stood there ... staring. People must have laughed if they saw me ...

... What's that? The clock? Oh, madam, you should have stopped me! Let me cover up your feet. I always cover up my lady's feet, every night. And she always says, 'Goodnight, Ellen. Sleep well and wake up early!'

... Oh dear, I sometimes think ... whatever would I do if anything happened ... But thinking's no good, is it, madam? Thinking won't help. When I find myself doing that, I say to myself, 'Come along, Ellen! Stop it this moment, my girl! Stop that silly thinking ... !'

GLOSSARY

balcony a platform built on the outside wall of a building, which you can walk on to from an upstairs room

ball a party for dancing, with an organized programme

bark *(v)* to make a short, sharp sound (the noise made by dogs)

blush to turn red with shyness or embarrassment

brooch a pretty pin (of silver, gold, etc.) worn on a woman's clothes

carriage a part of a railway train where passengers sit

cart a vehicle with two wheels, pulled by a horse

class (social class) a group of people at the same social level (rich or poor, educated or uneducated, etc.)

cottage a small simple house (usually in the country, by the sea, etc.)

creek a small river

curling tongs a metal tool that is heated and used to make straight hair curl

daffodil a yellow spring flower

darling a word for a person or thing that is much loved

dawn the first light of a new day

doll a toy that looks like a person, for children to play with

donkey an animal like a horse, but with short legs and long ears

fair *(n)* a big market, held in a field, with rides and games

fellow *(informal)* a man or boy

feuille d'album the French for 'a page from an album' (perhaps a book of family photographs)

fill in to write details (e.g. your name, address, age) on a form, etc.

form *(n)* an official piece of paper containing questions and places to write in the answers

Frau the German word for 'Mrs' (a married woman)

gentleman a man who is polite and who behaves well to other people

governess a woman employed to teach young children in their home

hairdresser a person whose job is to cut, wash, curl, etc. people's hair

hair-do *(informal)* the way of arranging a woman's hair

handkerchief a square piece of cloth which can be worn around the neck or over the hair

ice-cream a sweet frozen food

karaka tree a New Zealand tree with orange fruit

kid *(informal)* a child or young person

landlady a woman who rents rooms in her house to paying guests

lantern a light in a metal and glass case, often used outside

lawn an area of short neat grass in a garden or park

lily a plant with large white or coloured sweet-smelling flowers

Mademoiselle the French word for 'Miss' (an unmarried woman)

maid a girl or woman who works as a servant

marquee a large tent used for social events in gardens

mate *(informal)* a friend or companion (usually a man)

old man *(slang)* somebody's husband

paddock a small field where horses are kept

part *(n)* the character played by an actor in a play or film

partner one of two people dancing together

prayers words spoken to God

rare *(adj)* unusual, not often happening or seen

saddle a leather seat for a rider on a horse

scone a soft flat cake, often eaten with butter

shearing cutting the wool off a sheep

show *(n)* a play in a theatre, often with singing and dancing

stage a raised platform in a theatre or hall, for actors or
  musicians
stick (past tense **stuck**) to join or fasten together, or to push
  something into something else
store *(n)* a shop
strawberry a soft juicy red fruit
studio a room where an artist paints, and may also live
swing (past tense **swung**) to move backwards and forwards
topping *(informal, not used today)* wonderful
tram an electric bus which runs on rails, seen in many big cities
weary *(adj)* sad and tired
West End an area of London with many theatres, restaurants, etc.

NOTE ON USAGE

I seen a non-standard form of 'I saw' (used by uneducated
  speakers)

The Garden Party
and Other Stories

ACTIVITIES

Before Reading

1 Read the story titles on the contents page. In which story do you think you will find these people? (Use the glossary to help you.)

   1 A woman whose job is teaching children in their own home.
   2 Some children who are given a wonderful present.
   3 A rich family who enjoy inviting guests to their home.
   4 A shy country girl who has never been to a dance before.
   5 A woman who keeps a shop in a lonely country area.
   6 A poor woman who spends her life serving other people.
   7 Someone who is trying to find work in the theatre or films.

2 Read the back cover and the introduction on the first page. Which of these things do you think you will find in the stories?

a fatal accident	madness	a dangerous journey
a murder	a wedding	a forbidden love affair
a jealous lover	happy endings	family disagreements
money worries	a new baby	an undiscovered crime

3 What do you think might happen to the people described in the introduction? Choose Y (yes) or N (no) for each sentence.

   1 The girl will enjoy her first dance. Y/N
   2 The woman will leave the lonely farm. Y/N
   3 The artist will fall in love. Y/N
   4 The desperate search for work will be successful. Y/N
   5 The train journey across Europe will end in disaster. Y/N
   6 The cruel children will be punished. Y/N

While Reading

Read the first story, *Feuille d'Album*. Here are some untrue sentences about it. Change them into true sentences.

1 Ian French was a journalist, who lived in a dirty, untidy flat.
2 He was confident and talkative, but women didn't find him very interesting.
3 He fell in love with a girl when he met her in the street.
4 When he first spoke to her, he gave her some flowers.

Read *The Doll's House*, and answer these questions.

1 What did Kezia like best about the doll's house, and why?
2 Why did the other children never speak to the little Kelveys?
3 In what ways were the little Kelveys seen as different?
4 Why do you think the other girls were so cruel to the Kelveys?
5 Why do you think Kezia invited the Kelveys into the garden to see the doll's house?

Read *The Garden Party* up to the end of page 23. Can you guess how the story continues? Choose one of these ideas.

1 The Sheridans cancel the party.
2 They have the party, but there is no music.
3 They have the party a week later.
4 They have the party, and afterwards they send some left-over food to the family of the dead man.
5 They have the party, and collect some money for the family of the dead man.

Read *Pictures*. Who said this, and to whom? What were they talking about?

1 'I'm sick and tired of it all, and I've had enough.'
2 'Give it back to me at once, you bad, wicked woman.'
3 'I've got enough money for that.'
4 'They had to be young and able to kick their legs up a bit.'
5 'I could just go in and sit there and have a coffee, that's all.'
6 'What's yours?'

Read *The Little Governess*. Then put these parts of sentences into the correct order, to make a paragraph of four sentences. Begin with number 3.

1 and accepted his offer to show her round Munich for the day.
2 She made her escape and hurried back to the hotel,
3 The little governess did not know the ways of the world,
4 so she was soon talking to him happily,
5 When an old man got into her carriage on the train,
6 but found she had missed her meeting with her new employer.
7 However, the old man was not as pleasant as he seemed,
8 and her first journey abroad ended in disaster.
9 which she finally realized when he demanded a kiss.
10 he was polite to her and kind in a grandfatherly way,

Read *Her First Ball*, and answer these questions.

1 Why had Leila never been to a ball before?
2 What did the young men talk about while dancing with Leila?
3 How long had the fat man been going to balls?
4 What did he tell Leila that made her want to cry?
5 What made her feel happy again?

Read *The Woman at the Store*. Are these sentences true (T) or false (F)? Change the false ones into true sentences.

1 The woman at the store had not changed at all in four years.
2 She lived alone, with only a child and a dog for company.
3 The woman said that her husband was dead and buried.
4 Because of the storm, the travellers slept in the building.
5 The kid talked about her mother shooting and burying a man.

Read *Millie*. Then put these parts of sentences into the correct order, to make a paragraph of four sentences. Begin with number 5.

1 that she wanted to help him escape.
2 As Millie watched them rushing after Harrison,
3 she saw that he was just a young and terrified boy.
4 and she screamed at the men to shoot him down.
5 When Millie found Harrison behind the wood pile,
6 but her heart was so full of pity for him
7 the men sleeping in the house were woken by a barking dog.
8 She knew that he had killed a man,
9 all her feelings of pity for the boy disappeared,
10 However, when he took Sid's horse in the night and rode away,

Read *The Lady's Maid*, and answer these questions.

Why

1 ... was it no trouble for Ellen to bring the visitor a cup of tea?
2 ... did her grandfather burn her hand when she was young?
3 ... couldn't she remember ever feeling she was a child?
4 ... did she tell Harry she wouldn't marry him?
5 ... did she worry about the future?

After Reading

1 **Perhaps this is what some of the characters in the stories were thinking. Which characters were they (there is one from each story), and what was happening in the story at that moment?**

  1 'Well, here's a new one! The first time she's been out, I imagine.
    Young, shy, and rather pretty. But it won't last long – they start
    to look old so quickly. Now, where's my programme . . . ?'

  2 'Where does the child get her strange ideas from? Sometimes I
    don't understand her at all. How could we disappoint all our
    friends? It really is too selfish of her. Now, what can I do to take
    her mind off this nonsense? Ah, I know . . .'

  3 'Nearly home now. What a lovely evening! Spring's on its way –
    I do love the spring flowers. Now, where's my key . . . Who's
    that? Someone's following me! Who can it be?'

  4 'She shouldn't leave me in here with these two! They're
    laughing at me, and they're horrible! I don't care what she said.
    I'm going to draw it. Then she'll be sorry!'

  5 'I mean it! I'll give her one more day, and if she doesn't bring me
    the money, she'll have to go. She's been lying to me for long
    enough, and I can get good money for that room . . .'

  6 'Is she really going to leave me after all these years? How will I
    manage without her? I'm just not used to doing things for
    myself – I don't know how to!'

7 'Sweet little thing . . . Wonderful hair! And so innocent! No idea of how the world goes . . . Well, she's cost me enough for one day. Time to invite her to pay a little visit . . .'

8 'What's that I can hear in the garden? Whose voices are those? No, I don't believe it! How many times have I told her she must never, never . . .'

9 'What's she doing in there? Has she really gone to get me something to eat? Does she know who I am? Oh God! What will they do to me when they get back . . . ?'

2 **At the end of *Feuille d'Album*, how did the conversation between Ian French and the girl continue? Complete their conversation (use as many words as you like).**

IAN: Excuse me, Mademoiselle, you dropped this.

GIRL: That's not mine. I haven't dropped any eggs!

IAN: _____

GIRL: Well, thank you. That's very kind of you. But who are you?

IAN: _____

GIRL: Oh yes, the tall building. So you're an artist, are you? What kind of pictures do you paint?

IAN: _____

GIRL: You'd like to paint me? Oh no, I don't think so . . .

IAN: _____

GIRL: She's not my mother, she's my grandmother.

IAN: _____

GIRL: Visit us? Oh, I don't know. We don't have many visitors, but . . . well, yes, all right.

IAN: _____

GIRL: Come tomorrow if you like. Yes, come for coffee at ten.

3   Here is Lil, from *The Doll's House*, now grown up. She is talking
    about her childhood, when she was at school. Choose a suitable
    word for each gap.

    'We had no money as kids, _____ what was much worse, we had
    _____ friends. The other girls at school _____ not even allowed to
    talk to _____. There was one girl, Kezia Burnell, _____ tried to be
    nice to us. _____ Burnells had this great big doll's _____ in their
    garden – the other girls _____ stop talking about it. Anyway, Kezia
    _____ us in to see it, but _____ were only in the garden for _____
    minute before that old cat, Kezia's _____, came and threw us out!
    It _____ our Else so happy, though. She _____ ever spoke when she
    was a _____, just followed me around, holding my _____. But I
    always knew when she _____ something, and she really wanted to
    _____ that doll's house, with its little _____, just like a real one!
    She _____ forgot it, poor Else!'

4   Here are some remarks made by characters in *The Garden Party*.
    Who said them, and what do the remarks tell you about the people
    and their opinions?

    1   'You won't bring a drunk workman back to life by stopping a
        party.'
    2   'I say, Laura, take a look at my coat, can you, before this
        afternoon? I think it needs ironing.'
    3   'No, wait, take some lilies too. These lilies will seem like
        something really special to people of that kind.'
    4   'But we can't possibly have a garden party with a man dead just
        outside the front gate.'
    5   'It was a horrible thing, though. The fellow was married, too.
        . . . Leaves a wife and a whole crowd of kids, they say.'

5 These three women, from *Pictures*, *Her First Ball*, and *The Lady's Maid*, had very different lives. Choose the right notes for each woman, and use them to write short descriptions.

Miss Moss / Leila / The lady's maid

1 an unloved child / began work aged thirteen / chose not to marry / cheerful / unselfish / anxious about the future
2 a singer and actress / getting old and fat / no work / lived alone / no friends or family / friendly / brave / not lucky
3 young and pretty / protected life / parents and relations / no need to earn a living / shy / innocent / easily upset

Now imagine these three women five years after the end of the stories. What do you think their lives are like now?

6 What do you think happened to the little governess after the end of the story? Choose one of these ideas, and continue the story.

1 Suddenly, the little governess came face to face with an angry-looking lady, who stared at her. 'Miss!' the lady said. 'Are you the English governess I expected to meet here?' The little governess blushed. 'Oh, Frau Arnholdt!' she cried. 'Please . . .'

2 With tears running down her face, the little governess ran to the hotel manager's office, and rushed in without knocking. 'Oh, please help me!' she cried. 'Something terrible has happened to me!' The manager stared at her in surprise . . .

3 Too upset to think properly, the little governess turned and ran out of the hotel into the street. Where could she go? It was almost dark, and she was alone in a strange city, without a single friend, and with very little money . . .

7 After the end of *The Woman at the Store*, Jo catches up with his friends. Put his conversation with Jim in the correct order, and write in the speakers' names. Jim speaks first (number 3).

1 _____ 'She's a killer, Jo! That husband of hers – he's not away shearing, he's dead and buried. And she shot him!'

2 _____ 'That kid's sick in the head! Any fool could see that!'

3 _____ 'Jo! Good to see you safe and well!'

4 _____ 'Shut it, mate! Her old man's gone shearing, and she gets a bit lonely – end of story! Come on, let's get moving . . .'

5 _____ 'Oh, is that so? Listen, Jo, I've got something to tell you about that woman.'

6 _____ 'Shot him? You're crazy! Where did you get that idea?'

7 _____ 'She may be sick, Jo, but she saw what her Ma did, I'm sure of it. And that woman shot and buried her husband!'

8 _____ 'I've been well looked after! She's a good woman, Jim – knows how to take care of a man!'

9 _____ 'The kid told us – well, not in words, she drew a picture.'

10 _____ 'What about her? Why are you looking at me like that?'

8 Complete this newspaper report of the story told in *Millie* (use as many words as you like). What do you think Millie might have said? Was she pleased, or angry, at Harrison's death?

KILLER OF MR WILLIAMSON CAUGHT AND SHOT

Harrison, the young Englishman who is believed _____, disappeared after the shooting. However, last night he _____. Sid Evans and his men were woken _____. They shouted at Harrison to stop, but _____ so _____.

Mrs Millie Evans, wife of Sid Evans, told us, 'What happened to Mr Williamson was _____. I _____.'

9 Do you think the characters in these stories made the right decisions and did the right things? If not, what do you think they should have done?

1 Ian French found an unusual way to meet the girl he had seen.
2 Kezia invited the Kelveys to see the doll's house, although her mother and aunt had told her not to do that.
3 The Sheridans had their party, although a man from the cottages near their house had just been killed.
4 Miss Moss followed the large gentleman out of the café.
5 The little governess accepted an invitation from a man she had met on the train.
6 The fat man told Leila what he was thinking.
7 The travellers rode away and left the child with the woman at the store.
8 Millie didn't tell Sid that she had seen and helped Harrison.
9 Ellen decided not to marry Harry so that she could look after 'her lady'.

10 Here are some different titles for the stories. Which stories could they go with? Can you suggest some more titles of your own?

A Shocking Death   Gone Shearing
The Comfort of Lilies   The Little Lamp
A Life of Service   A Kind Heart
Springtime in the Heart   The Dancing Partner
A Woman in Trouble   Talking to Strangers

11 Which story did you enjoy most? Which character did you feel most sympathetic towards, or find most interesting? Why?

ABOUT THE AUTHOR

Katherine Mansfield was born Kathleen Mansfield Beauchamp in Wellington, New Zealand, in 1888. Her father was a wealthy banker, and Katherine, with her brother and three sisters, had a comfortable life as a child. At the age of fourteen she was sent to London to finish her education, and she spent most of the rest of her short life in Europe.

From an early age, Katherine felt different from her family. She studied music for a time, and was already writing stories while still at school. She spent her life among writers and artists, and her friends included some famous authors of the time, such as D. H. Lawrence and Virginia Woolf.

Although she married twice, Katherine never lived an ordinary family life. From the age of twenty, she suffered from a serious disease, and in search of better health, she spent part of every year in France and Switzerland. She wrote a large number of short stories, even though it was often difficult for her to find the strength and peace she needed in order to write. She died in France in 1923, aged only thirty-five.

Her first book, *In a German Pension*, appeared in 1911, followed by *Prelude* in 1916, *Bliss* in 1921, and *The Garden Party and Other Stories* in 1922. Two more books of stories, her letters, and her journal were published after her death.

She is considered to be one of the finest writers of her time, and has often been compared to the Russian writer Chekhov. Her sensitive, delicate stories take the reader straight into the lives of her characters, who are often women struggling to survive in an unfriendly world.

ABOUT BOOKWORMS

OXFORD BOOKWORMS LIBRARY
Classics • True Stories • Fantasy & Horror • Human Interest
Crime & Mystery • Thriller & Adventure

The OXFORD BOOKWORMS LIBRARY offers a wide range of original and adapted stories, both classic and modern, which take learners from elementary to advanced level through six carefully graded language stages:

Stage 1 (400 headwords)	**Stage 4** (1400 headwords)
Stage 2 (700 headwords)	**Stage 5** (1800 headwords)
Stage 3 (1000 headwords)	**Stage 6** (2500 headwords)

More than fifty titles are also available on cassette, and there are many titles at Stages 1 to 4 which are specially recommended for younger learners. In addition to the introductions and activities in each Bookworm, resource material includes photocopiable test worksheets and Teacher's Handbooks, which contain advice on running a class library and using cassettes, and the answers for the activities in the books.

Several other series are linked to the OXFORD BOOKWORMS LIBRARY. They range from highly illustrated readers for young learners, to playscripts, non-fiction readers, and unsimplified texts for advanced learners.

Oxford Bookworms Starters	*Oxford Bookworms Factfiles*
Oxford Bookworms Playscripts	*Oxford Bookworms Collection*

Details of these series and a full list of all titles in the OXFORD BOOKWORMS LIBRARY can be found in the *Oxford English* catalogues. A selection of titles from the OXFORD BOOKWORMS LIBRARY can be found on the next pages.

Heat and Dust

RUTH PRAWER JHABVALA

Retold by Clare West

Heat and dust – these simple, terrible words describe the Indian summer. Year after year, endlessly, it is the same. And everyone who experiences this heat and dust is changed for ever.

We often say, in these modern times, that sexual relationships have changed, for better or for worse. But in this book we see that things have not changed. Whether we look back sixty years, or a hundred and sixty, we see that it is not things that change, but people. And, in the heat and dust of an Indian summer, even people are not very different after all.

The Bride Price

BUCHI EMECHETA

Retold by Rosemary Border

When her father dies, Aku-nna and her young brother have no one to look after them. They are welcomed by their uncle because of Aku-nna's 'bride price' – the money that her future husband will pay for her.

In her new, strange home one man is kind to her and teaches her to become a woman. Soon they are in love, although everyone says he is not a suitable husband for her. The more the world tries to separate them, the more they are drawn together – until, finally, something has to break.

Wuthering Heights

EMILY BRONTË

Retold by Clare West

The wind is strong on the Yorkshire moors. There are few trees, and fewer houses, to block its path. There is one house, however, that does not hide from the wind. It stands out from the hill and challenges the wind to do its worst. The house is called Wuthering Heights.

When Mr Earnshaw brings a strange, small, dark child back home to Wuthering Heights, it seems he has opened his doors to trouble. He has invited in something that, like the wind, is safer kept out of the house.

Far from the Madding Crowd

THOMAS HARDY

Retold by Clare West

Bathsheba Everdene is young, proud, and beautiful. She is an independent woman and can marry any man she chooses – if she chooses. In fact, she likes her independence, and she likes fighting her own battles in a man's world.

But it is never wise to ignore the power of love. There are three men who would very much like to marry Bathsheba. When she falls in love with one of them, she soon wishes she had kept her independence. She learns that love brings misery, pain, and violent passions that can destroy lives . . .

Jeeves and Friends

P. G. WODEHOUSE

Retold by Clare West

What on earth would Bertie Wooster do without Jeeves, his valet? Jeeves is calm, tactful, resourceful, and has the answer to every problem. Bertie, a pleasant young man but a bit short of brains, turns to Jeeves every time he gets into trouble. And Bertie is *always* in trouble.

These six stories include the most famous of P. G. Wodehouse's memorable characters. There are three stories about Bertie and Jeeves, and three about Lord Emsworth, who, like Bertie, is often in trouble, battling with his fierce sister Lady Constance, and his even fiercer Scottish gardener, the red-bearded Angus McAllister . . .

Dublin People

MAEVE BINCHY

Retold by Jennifer Bassett

A young country girl comes to live and work in Dublin. Jo is determined to be modern and independent, and to have a wonderful time. But life in a big city is full of strange surprises for a shy country girl . . .

Gerry Moore is a man with a problem – alcohol. He knows he must give it up, and his family and friends watch nervously as he battles against it. But drink is a hard enemy to fight . . .

These stories by the Irish writer Maeve Binchy are full of affectionate humour and wit, and sometimes a little sadness.